Lecture Notes in Mathematics

Edited by A. Dold and B. Eckmann

509

David E. Blair

Contact Manifolds in Riemannian Geometry

Springer-Verlag
Berlin · Heidelberg · New York 1976

Author
David Ervin Blair
Department of Mathematics
Michigan State University
East Lansing, Michigan 48824
USA

Library of Congress Cataloging in Publication Data

Blair, David E 1940-
Contact manifolds in Riemannanian geometry.

(Lecture notes in mathematics ; 509)
"A slightly expanded version of the author's lectures at the University of Strasbourg and the University of Liverpool during the academic year 1974-75."
Bibliography: p.
Includes index.
1. Geometry, Riemannian. 2. Riemannian manifolds.
I. Title. II. Series: Lecture notes in mathematics (Berlin) ; 509.
QA3.L28 no. 509 [QA649] 510'.8s [516'.373] 76-3757

AMS Subject Classifications (1970): 53-02, 53C15, 53C25

ISBN 3-540-07626-3 Springer-Verlag Berlin · Heidelberg · New York
ISBN 0-387-07626-3 Springer-Verlag New York · Heidelberg · Berlin

Printed in Germany
Printing and binding: Beltz Offsetdruck, Hemsbach/Bergstr.

PREFACE

These notes are a slightly expanded version of the author's lectures at the University of Strasbourg and the University of Liverpool during the academic year 1974-75. As the subject of contact manifolds from the Riemannian point of view was not well known to the audiences, these lectures provided an introduction to the subject with some recent work given in the later lectures.

The two large classes of classical examples of contact manifolds are the principal circle bundles of the Boothby-Wang fibration including the odd-dimensional spheres and the tangent sphere bundles. The former class and its generalizations have received considerable attention as contact metric manifolds but the latter class has not. An attempt is made here to give some insight into the geometry of both these classes (chiefly in Chapters IV and VII).

The contact form on a contact manifold determines a distribution or subbundle, called the contact distribution, which is far from being integrable. In fact the maximum dimension of an integral submanifold is half the dimension of the contact distribution. This is discussed in Chapter III and Chapter V deals with recent work on the integral submanifolds of a Sasakian space form.

Finally in Chapter VI we prove the non-existence of a flat associated metric to a contact structure on a contact manifold of dimension ≥ 5. Thus while it is an open question as to whether or not the 5-dimensional torus carries a contact structure, such a structure could not have the flat metric on the torus as an associated metric.

The author expresses his appreciation to his colleagues at both the above mentioned universities and at Michigan State University for their helpful discussions and encouragement in this project and to Mrs. Glendora Milligan for her excellent work in typing the manuscript.

East Lansing
November 1975

David E. Blair

CONTENTS

CHAPTER I

CONTACT MANIFOLDS

1. Contact Manifolds.

A differentiable (C^∞) $(2n+1)$-dimensional manifold M^{2n+1} is said to be a contact manifold or to have a contact structure (in the restricted sense) if it carries a global differential 1-form η such that

$$\eta\wedge(d\eta)^n \neq 0$$

everywhere on M^{2n+1} where the exponent denotes the nth exterior power. We call η a contact form. Note, in particular, that a contact manifold in this sense is orientable.

Before defining a contact manifold in the wider sense, we recall the classical theorem of Darboux (see e.g. Cartan [14], Sternberg [65]).

Theorem. Let ω be a 1-form on a differentiable manifold M^n and suppose that $\omega\wedge(d\omega)^p \neq 0$ and $(d\omega)^{p+1} \equiv 0$ on M^n. Then about every point there exists a coordinate system $(x^1,\cdots, x^p, y^1,\cdots, y^{n-p})$ such that $\omega = dy^{p+1} - \sum_{i=1}^{p} y^i dx^i$.

Thus, about every point of a contact manifold M^{2n+1} there exists coordinates (x^i,y^i,z), $i = 1,\cdots, n$ such that

$$\eta = dz - \sum_{i=1}^{n} y^i dx^i .$$

Now let (x^i, y^i, z), $i = 1, \cdots, n$ be cartesian coordinates on $\mathbb{R}^{2n+1}$ and let $\eta = dz - \sum_{i=1}^{n} y^i dx^i$. A diffeomorphism ($C^\infty$) $f: U \to U'$ where U and U' are open subsets of $\mathbb{R}^{2n+1}$ is called a <u>contact transformation</u> if the pullback $f^*\eta = \tau\eta$ for some nonvanishing function τ on U. The collection Γ of all such contact transformations is closed under composition, formation of inverses, and restriction to open subsets, and hence forms a pseudogroup.

A differentiable manifold M^{2n+1} is said to be a <u>contact manifold in the wider sense</u> if there exists an open covering $\{U_\alpha\}$ of M^{2n+1} with homeomorphisms $f_\alpha: U_\alpha \to V_\alpha \subset \mathbb{R}^{2n+1}$ such that $f_\alpha \circ f_\beta^{-1} \in \Gamma$ for all α and β for which $f_\alpha \circ f_\beta^{-1}$ is defined. Two such atlases $\{U_\alpha, f_\alpha\}$ and $\{U'_\gamma, f'_\gamma\}$ are said to be equivalent if $f'_\gamma \circ f_\alpha^{-1} \in \Gamma$ whenever defined. An equivalence class is then called a <u>contact structure in the wider sense</u> on M^{2n+1}.

Thus from the Darboux Theorem we see that a contact manifold (in the restricted sense) is a contact manifold in the wider sense.

More geometrically a contact structure in the wider sense may be thought of as follows (see e.g. Martinet [40]). Let M^n be a C^∞ manifold. A <u>hyperplane field</u> on M^n is an $(n-1)$-dimensional subbundle D of the tangent bundle TM^n. Such a field may be defined as a section σ of the bundle $PM^n \to M^n$ whose fibre P_mM^n at $m \in M^n$ is the projective space of lines in the cotangent space $T^*_m M^n$, the hyperplane $\sigma_m \subset T_m M^n$ being defined by the line $\sigma(m) \subset T^*_m M^n$.

Locally a hyperplane field may be defined by equations $\omega = 0$, where ω is a 1-form without zeros. Clearly two 1-forms ω and ω' defining the same field satisfy $\omega' = \tau\omega$ for some non-vanishing function τ and a hyperplane field is often called a Pfaffian equation. If σ is a hyperplane field and ω a 1-form defining σ in a neighborhood of $m \in M^n$, the integer $C(M) = \max\{2p+1 \mid \omega\wedge(d\omega)^p \neq 0$ at $m\}$ is called the class of σ at m. A point $m \in M^n$ is called a singular point of σ if $C(m) < n$.

Thus a contact structure in the wider sense is a hyperplane field σ without singularities, that is, σ has maximal class everywhere and hence, in particular, the manifold is odd-dimensional.

If now σ is a contact structure in the wider sense defined locally by contact forms η_α on neighborhoods U_α where $\{U_\alpha\}$ is an open covering of M^{2n+1}, the fibres D_m of the corresponding subbundle D of TM^{2n+1} are given by

$$D_m = \{X \in T_m M^{2n+1} \mid \eta_\alpha(X) = 0\}$$

for $m \in U_\alpha$. Recall that a vector bundle over a manifold with fibre $\mathbb{R}^p$ is said to be orientable if the structural group of its associated principal bundle with group $GL(p,\mathbb{R})$ can be reduced to $GL^+(p,\mathbb{R})$, the subgroup of matrices with positive determinant (see e.g. Kobayashi and Nomizu [33]).

Now since η_α is a contact form on U_α, $(d\eta_\alpha)^n \neq 0$ on D_m and hence $d\eta_\alpha$ is a non-degenerate skew-symmetric bilinear form on D_m uniquely determined to within a non-zero multiple. If now $\eta_\alpha = \tau_{\alpha\beta}\eta_\beta$ on $U_\alpha \cap U_\beta$, $\eta_\alpha \wedge (d\eta_\alpha)^n = \tau_{\alpha\beta}^{n+1}(\eta_\beta \wedge (d\eta_\beta)^n)$ and $\tau_{\alpha\beta}^{n+1}$ is just the Jacobian of the coordinate transformation. Thus if M^{2n+1} is orientable and n even, $\tau_{\alpha\beta}$ is positive and the vector bundle D is orientable.

We now show that if a contact manifold in the wider sense M^{2n+1} is orientable and n is even, then it is a contact manifold (in the restricted sense), (Gray [23], see also Kobayashi [32], Stong [66]).

Theorem. Let M^{2n+1} be an orientable contact manifold in the wider sense with n even. Then M^{2n+1} is a contact manifold.

Proof. Since M^{2n+1} is orientable, its tangent bundle is orientable as a vector bundle and hence, since D is orientable, the quotient bundle TM^{2n+1}/D is an orientable real line bundle. Thus, in particular, TM^{2n+1}/D admits a cross section S without zeros. On the other hand, η_α defines a local cross section S_α over U_α by the equation $\eta_\alpha(S_\alpha) = 1$, and hence, $S_\alpha = h_\alpha S$ where the h_α's are non-vanishing functions of the same sign. Now defining η by $\eta = h_\alpha\eta_\alpha$ on U_α we obtain a global 1-form η such tha $\eta \wedge (d\eta)^n \neq 0$.

If n is odd, Gray [23] has shown that M^{2n+1} is orientable, but now D might not be. If M^{2n+1} is a contact manifold in the wider sense which is not a contact manifold in our sense, let $\tilde{M}^{2n+1}$ be a double covering of M^{2n+1}. Then the contact structure on M^{2n+1} lifts to one on $\tilde{M}^{2n+1}$ whose subbundle D is orientable. Applying the above argument $\tilde{M}^{2n+1}$ admits a contact form (Sasaki [59]). Thus, the two notions of contact structure are closely related and except for example 2.B. below, we will regard a contact manifold as defined at the outset.

We close this section with some additional definitions that will be needed later. Let M^{2n+1} be a contact manifold, then the subbundle or 2n-dimensional distribution D on M^{2n+1} is called the <u>contact distribution</u>. The condition $\eta\wedge(d\eta)^n \neq 0$ implies that D is not integrable; if fact loosely speaking D is as far from being integrable as possible. This phenomenon is discussed more fully in Chapters III and V.

As we have seen, the orientability of M^{2n+1} and D implies that the line bundle TM^{2n+1}/D admits a cross section S on which $\eta(S) = 1$. Thus M^{2n+1} admits a global non-vanishing vector field, denoted by ξ, such that

$$\eta(\xi) = 1$$

and

$$d\eta(\xi,X) = 0$$

for all vector fields X on M^{2n+1}. Clearly ξ determines a 1-dimensional distribution complementary to D and we call ξ the <u>characteristic vector field</u> of the contact structure.

Two basic properties of ξ are the invariance of η and $d\eta$ under its 1-parameter group. Letting $\mathfrak{L}$ denote Lie differentiation, $\mathfrak{L}_\xi \eta = 0$ and $\mathfrak{L}_\xi d\eta = 0$ follow immediately from $\eta(\xi) = 1$ and $d\eta(\xi,X) = 0$ using the formula $\mathfrak{L}_\xi = d \circ \iota(\xi) + \iota(\xi) \circ d$ where $\iota(\xi)$ is the interior product by ξ.

Finally we say that a contact structure is <u>regular</u> if ξ is a regular vector field on M^{2n+1}, that is every point $m \in M^{2n+1}$ has a cubical coordinate neighborhood U such that the integral curves of ξ passing through U pass through the neighborhood only once.

2. Examples

A. $\mathbb{R}^{2n+1}$. We have already seen that $\eta = dz - \sum_{i=1}^{n} y^i dx^i$ is a contact structure on $\mathbb{R}^{2n+1}$, (x^i, y^i, z) being cartesian coordinates. The vector field ξ is just $\partial/\partial z$ and the contact distribution D is spanned by

$$X_i = \frac{\partial}{\partial x^i} + y^i \frac{\partial}{\partial z}, \quad X_{n+i} = \frac{\partial}{\partial y^i}$$

$i = 1, \cdots, n$.

B. $\mathbb{R}^{n+1} \times P\mathbb{R}^n$. We now give an example of a contact manifold in the wider sense which is not a contact manifold in our sense (Gray [23]). Consider $\mathbb{R}^{n+1}$ with coordinates $(x^1, \cdots, x^{n+1})$ and real projective space $P\mathbb{R}^n$ with homogeneous coordinates $(t_1, \cdots, t_{n+1})$ and let $M^{2n+1} = \mathbb{R}^{n+1} \times P\mathbb{R}^n$. The subsets $\{U_i\}$ $i = 1, \cdots, n+1$ defined by $t_i \neq 0$, form an open covering of M^{2n+1} by coordinate neighborhoods. On U_i define a 1-form η_i by $\eta_i = \frac{1}{t_i} \sum_{j=1}^{n+1} t_j dx^j$, then $\eta_i \wedge (d\eta_i)^n \neq 0$ and $\eta_i = \frac{t_j}{t_i} \eta_j$. Thus, M^{2n+1} has a contact structure in the wider sense, but for n even, M^{2n+1} is non-orientable and hence cannot carry a global contact form.

C. T^3. In [40] J. Martinet proved that every compact orientable 3-manifold carries a contact structure. Here we will just give explicitly a contact structure on the 3-dimensional torus T^3.

First consider $\mathbb{R}^3$ with coordinates (x^1, x^2, x^3) and define η by $\eta = \cos x^3 dx^1 + \sin x^3 dx^2$. Then $\eta \wedge d\eta = - dx^1 \wedge dx^2 \wedge dx^3$. Now let Γ be the group of transla tions of $\mathbb{R}^3$ with generators $\{x^i \to x^i + 2\pi,\ i = 1,2,3\}$; then $T^3 = \mathbb{R}^3/\Gamma$ is a torus which clearly carries the contac structure η.

The characteristic vector field ξ of this contact structure is the vector field $\cos x^3 \frac{\partial}{\partial x^1} + \sin x^3 \frac{\partial}{\partial x^2}$. Th integral curve of ξ through $(0,0,\frac{\pi}{3})$ is given by $x^1 = \frac{t}{2}$ $x^2 = \frac{\sqrt{3}}{2} t$, $x^3 = \frac{\pi}{3}$. Therefore ξ induces an irrational flow on the 2-dimensional torus $x^3 = \frac{\pi}{3}$ and hence the conta structure η on T^3 is not regular.

We shall show in section IV.4. that T^3 cannot carry a regular contact structure. For the same reason the 5-dimensional torus also cannot carry a regular contact structure, although it is still an open question whether or not T^5 carries any contact structure. By the theorem of section VI.1. T^5 could not have a flat Riemannian metric as an associated metric.

D. $M^{2n+1} \subset \mathbb{R}^{2n+2}$ with $T_m M^{2n+1} \cap \{0\} = \emptyset$, S^{2n+1}, $P\mathbb{R}^{2n+1}$.

Here we prove the following theorem (Gray [23]).

Theorem. Let $\iota : M^{2n+1} \to \mathbb{R}^{2n+2}$ be a smooth hypersurface immersed in $\mathbb{R}^{2n+2}$ and suppose that no tangent space of M^{2n+1} contains the origin of $\mathbb{R}^{2n+2}$, then M^{2n+1} has a contact structure.

Proof. Let $(x^1,\cdots, x^{2n+2})$ be cartesian coordinates on $\mathbb{R}^{2n+2}$ and consider the 1-form

$$\alpha = x^1dx^2 - x^2dx^1 + \cdots + x^{2n+1}dx^{2n+2} - x^{2n+2}dx^{2n+1} .$$

Let $V_1,\cdots, V_{2n+1}$ be $2n+1$ linearly independent vectors at a point $x_0 = (x_0^1,\ldots, x_0^{2n+2})$ and define a vector W at x_0 with components

$$W^A = *dx^A(V_1,\cdots, V_{2n+1})$$

where $*$ is the Hodge star operator of the Euclidean metric on $\mathbb{R}^{2n+2}$. Then W is normal to the hyperplane spanned by $V_1,\cdots, V_{2n+1}$. Now regard x_0 as a vector with components x_0^A, then

$$(\alpha \wedge (d\alpha)^n)(V_1,\cdots, V_{2n+1}) = x_0 \cdot W.$$

Thus, if no tangent space of M^{2n+1} regarded as a hyperplane in $\mathbb{R}^{2n+2}$ contains the origin, then $\eta = \iota^*\alpha$ is a contact form on M^{2n+1}.

As a special case we see that an odd-dimensional sphere S^{2n+1} carries a contact structure. Moreover α is invariant under reflection through the origin, $(x^1,\cdots, x^{2n+2}) \to (-x^1,\cdots, -x^{2n+2})$ and hence the real projective space $P\mathbb{R}^{2n+1}$ is also a contact manifold.

Similarly consider the 1-form $\beta = \sum_{i=1}^{n+1} x^i dx^{n+1+i}$ and denote by $\mathbb{R}_1^{n+1}$ and $\mathbb{R}_2^{n+1}$ the subspaces defined by $x^i = 0$ and $x^{n+1+i} = 0$ respectively, $i = 1,\cdots, n+1$.

Then β induces a contact form on M^{2n+1} if and only if $M^{2n+1} \cap \mathbb{R}_1^{n+1} = \emptyset$, $M^{2n+1} \cap \mathbb{R}_2^{n+1}$ is an n-dimensional submanifold and no tangent space of $M^{2n+1} \cap \mathbb{R}_2^{n+1}$ in $\mathbb{R}_2^{n+1}$ contains the origin of $\mathbb{R}_2^{n+1}$.

E. T_1^*M, T_1M. We shall show that the cotangent sphere bundle and the tangent sphere bundle of a Riemannian manifold are contact manifolds (see e.g. Reeb [54], Sasaki [57]). Let M be an $(n+1)$-dimensional Riemannian manifold and T^*M its cotangent bundle. Let $(x^1, \cdots, x^{n+1})$ be local coordinates on a neighborhood U of M and $(p^1, \cdots, p^{n+1})$ coordinates on the fibres over U defined with respect to the dx^i's. If $\pi\colon T^*M \to M$ is the projection map, then p^i and $q^i = x^i \circ \pi$, $i = 1, \cdots, n+1$ are local coordinates on T^*M. On T^*M consider the 1-form β defined by $\beta = \sum_{i=1}^{n+1} p^i dq^i$ on a coordinate neighborhood. Now the bundle T_1^*M of unit cotangent vectors has empty intersection with the zero section of T^*M, its intersection with any fibre of T^*M is an n-dimensional sphere and no tangent space to this intersection contains the origin of the fibre; thus, just as in the discussion at the end of the last example, β induces a contact structure on the hypersurface T_1^*M.

Similarly one obtains a contact structure on the bundle T_1M of unit tangent vectors. In fact if G_{ij} denote the components of the metric with respect to the coordinates $(x^1, \cdots, x^{n+1})$, and $(v^1, \cdots, v^{n+1})$ are the fibre coordinates on TM, define β locally by $\beta = \sum_{i,j} G_{ij} v^j dq^i$ where $q^i = x^i \circ \pi$ and π is the projection map $\pi: TM \to M$.

F. $T^*M \times \mathbb{R}$. Let M be an n-dimensional manifold and T^*M its cotangent bundle. As in the previous example we can define a 1-form β on T^*M by the local expression $\sum_{i=1}^{n} p^i dq^i$. Let $M^{2n+1} = T^*M \times \mathbb{R}$, t the coordinate on $\mathbb{R}$ and $\mu: M^{2n+1} \to T^*M$ the projection to the first factor. Then $\eta = dt - \mu^*\beta$ is a contact form on M^{2n+1}.

The reader may recognize this example from classical mechanics. Suppose M is the 3N-dimensional configuration space of N particles in $\mathbb{R}^3$, then T^*M is the phase space of the dynamical system, the q^i's being the "generalized coordinates" and the p^i's the "conjugate momenta." Moreover a coordinate transformation on T^*M preserving Hamilton's equations also preserves $d\beta$ and is just a "homogeneous contact" or "canonical" transformation in the sense of physics.

The name contact (Berührungstransformation) seems to be due to S. Lie [37] and is natural in view of the simple example of Huygens' principle [28] (see MacLane [39]). Consider $\mathbb{R}^2$ with coordinates (q,r). The classical notion of a "line element" of $\mathbb{R}^2$ is a point together with a non-vertical line through the point. Thus, a line element may be regarded as a point in $\mathbb{R}^3 (= T^*\mathbb{R} \times \mathbb{R})$

determined by the point and the slope p of the line. Given a curve C in the plane without vertical tangents, its tangent lines determine a curve in $\mathbb{R}^3$ which is an integral curve of the contact distribution D determined by $\eta = dr - pdq$ on $\mathbb{R}^3$. If now C is a wave front, by Huygens' principle the new wave front C_t at time t is the envelope of the circular waves centered at all the points of C. Thus the transformation from C to C_t is a transformation of $\mathbb{R}^3$ into itself. Moreover, tangent wave fronts are mapped to tangent wave fronts, thus transformations mapping integral curves of D into integral curves of D are called contact transformations (cf. section III.1).

G. $\mathbb{C}^* \times \mathbb{R}$. This simple example is a polar coordinate version of Example A. We denote by $\mathbb{C}^*$ the set of non-zero complex numbers and by (r,θ) the usual polar coordinates. Then on $M^3 = \mathbb{C}^* \times \mathbb{R}$ we define η by $\eta = dz - r^2 d\theta$. Clearly η is a contact structure, $\xi = \partial/\partial z$ and the contact distribution D is spanned by $\partial/\partial r$ and $V = \frac{1}{r}\frac{\partial}{\partial\theta} + r\frac{\partial}{\partial z}$. We note that the integral curves of V are helices on the cylinders $r = \text{constant}$.

H. <u>The Boothby-Wang fibration</u>. We now give an important class of examples, namely certain principal circle bundles over symplectic manifolds whose symplectic 2-forms have integral period and conversely we will prove that a compact regular contact manifold is of this type.

It is well known (see e.g. Kobayashi [31]) that the set of principal circle bundles over a manifold M can be given a group structure isomorphic to the cohomology group $H^2(M,\mathbb{Z})$. Let M^{2n} be a symplectic manifold with fundamental 2-form Ω (i.e., $\Omega^n \neq 0, d\Omega = 0$) such that $[\Omega] \in H^2(M,\mathbb{Z})$ and $\pi: M^{2n+1} \to M^{2n}$ the corresponding principal circle bundle. If η' is a connection form on M^{2n+1}, then there exists a 2-form Ω' on M^{2n} such that $d\eta' = \pi^*\Omega'$. However, the characteristic class $[\Omega]$ is independent of the choice of connections (Kobayashi [31]), so that $[\Omega] = [\Omega']$. Thus, there exists a 1-form ω on M^{2n} such that $\Omega = \Omega' + d\omega$. Now $\pi^*\omega$ is horizontal and equivariant and S^1 is abelian so that $\pi^*\omega \circ R_{t*} = ad(t^{-1})\,\pi^*\omega = \pi^*\omega$ where R_t denotes right translation by $t \in S^1$ and R_{t*} its differential. Thus setting $\eta = \eta' + \pi^*\omega$, we have $\eta \circ R_{t*} = \eta$ and $d\eta = \pi^*\Omega$. Moreover if ξ is a vertical vector field such that $\eta'(\xi) = 1$, $\eta(\xi) = 1$ since $(\pi^*\omega)(\xi) = \omega(\pi_*\xi) = 0$. Now if at any point of M^{2n+1}, $X_1, \cdots, X_{2n}$ are linearly independent horizontal vectors, $(\eta\wedge(d\eta)^n)(\xi, X_1, \cdots, X_{2n})$ is non-zero. Thus, regarding the Lie algebra valued form η as a real-valued form, we see that η is a contact structure on M^{2n+1}.

Conversely we now give the theorem of Boothby and Wang [13].

<u>Theorem</u>. Let M^{2n+1} be a compact regular contact manifold with contact form η', then there exists a contact form $\eta = \tau\eta'$ for some non-vanishing function τ whose characteristic vector field ξ generates a free effective S^1 action on M^{2n+1}. Moreover M^{2n+1} is the bundle space of a principal circle bundle $\pi: M^{2n+1} \to M^{2n}$ over a symplectic manifold M^{2n} whose fundamental 2-form Ω determines an integral cocycle on M^{2n}; η is a connection form on the bundle with curvature form $d\eta = \pi^*\Omega$.

<u>Proof</u>. Since η' is regular, its characteristic vector field ξ' is a regular vector field and hence its maximal integral curves or orbits are closed subsets of M^{2n+1}; but M^{2n+1} is compact, so these integral curves are homeomorphic to circles. Moreover, as ξ' is regular, M^{2n+1} is a fibre bundle over a manifold M^{2n} (the set of maximal integral curves with the quotient topology) and we denote the projection by π.

Now let $f'_t: M^{2n+1} \to M^{2n+1}$ denote the 1-parameter group of diffeomorphisms generated by ξ' and define the period λ' of ξ' at $m \in M^{2n+1}$ by $\lambda'(m) = \inf\{t \mid t>0,\ f'_t(m) = m\}$. $\lambda'(m)$ is constant on each orbit and since there are no fixed points $(\xi' \neq 0)$ $\lambda'(m)$ is never zero. Also as the orbits are circles, $\lambda'(m)$ is not infinite. We will show that λ' is constant on all of M^{2n+1}. Our argument is due to Tanno [70] (see also (Sasaki [59]). Let h be a Riemannian metric on M^{2n} and let $g = \pi^*h + \eta' \otimes \eta'$. Then g is a Riemannian

metric on M^{2n+1} and ξ' is a unit Killing vector field with respect to g since $\eta'(\xi') = 1$ and $\mathfrak{L}_{\xi'}\eta' = 0$. If ∇ denotes the Riemannian connection of g,
$g(\nabla_{\xi'}\xi', X) = -g(\nabla_X\xi', \xi') = 0$ and hence the orbits of ξ' are geodesics. If γ is an orbit through m, let γ' be an orbit sufficiently near to γ that there exists a unique minimal geodesic from m to γ' meeting γ' orthogonally at m'. Then since f_t' is an isometry for all t, the image of the geodesic arc $\widehat{mm'}$ is orthogonal to γ and γ' for all t. Thus, as a point m on γ moves through one period along γ, the corresponding point on γ' moves through one period and hence λ' is constant on M^{2n+1}.

We now define η and ξ by $\eta = \frac{1}{\lambda'}\eta'$ and $\xi = \lambda'\xi'$. Since λ' is constant, ξ is the characteristic vector field of the contact form η. Moreover ξ has the same orbits as ξ' and its period function $\lambda \equiv 1$. Thus, the one parameter group $\{f_t\}$ of ξ depends only on the equivalence class mod 1 of t and the action so induced of S^1 is effective and free.

Since ξ is regular we can choose an open covering $\{U_\alpha\}$ of M^{2n+1} such that on U_α we have coordinates $(x^1, \cdots, x^{2n+1})$ with the integral curves of ξ being given by $x^1 = \text{const.}, \cdots, x^{2n} = \text{const.}$ Then $\{\pi(U_\alpha)\}$ is an open covering of M^{2n} and we can define local cross sections $S_\alpha: \pi(U_\alpha) \to M^{2n+1}$ with $\pi S_\alpha = \text{identity}$ by

$s_\alpha(x^1,\cdots, x^{2n}) = (x^1,\cdots, x^{2n},c)$ for some constant c, $x^1,\cdots, x^{2n}$ being regarded as the coordinates on $\pi(U_\alpha)$. Now the maps $\mu_\alpha\colon \pi(U_\alpha) \times S^1 \to M^{2n+1}$ defined by $\mu_\alpha(p,t) = f_t s_\alpha(p), p \in M^{2n}$ are coordinate functions for the bundle.

We have already seen that $\mathfrak{L}_\xi\eta = 0$ and $\mathfrak{L}_\xi d\eta = 0$, so that η and $d\eta$ are invariant under the action of S^1. Now take $A = d/dt$ as a basis of $\mathfrak{G} = \mathbb{R}$, the Lie algebra of S^1 and set $\tilde{\eta} = \eta A$ so that η may be regarded as a Lie algebra valued 1-form. For an element $B \in \mathfrak{G}$ we denote by B^* the induced vector field on M. Thus, $\tilde{\eta}$ is a connection form if $\tilde{\eta}(B^*) = B$ and $\tilde{\eta}$ is equivariant. Now $A^* = \xi$, so that $\tilde{\eta}(A^*) = A$. Moreover right translation by $t \in S^1$ is just f_t so that $R_t^*\tilde{\eta} = \tilde{\eta}$ by the invariance of η under the S^1 action. Thus, η (precisely $\tilde{\eta}$) is a connection form on M^{2n+1}.

If $\tilde{\Omega}$ is the curvature form of η on M^{2n+1}, then the structural equation is $d\eta = -\frac{1}{2}[\eta,\eta] + \tilde{\Omega} = \tilde{\Omega}$ since S^1 is abelian. On the other hand $d\eta$ is horizontal and invariant under the action of S^1, so there exists a 2-form Ω on M^{2n} such that $d\eta = \pi^*\Omega$. Now $\pi^* d\Omega = d\pi^*\Omega = d^2\eta = 0$, so that $d\Omega = 0$ and $\pi^*(\Omega)^n = (\pi^*\Omega)^n = (d\eta)^n \neq 0$ giving $\Omega^n \neq 0$. Therefore M^{2n} is symplectic.

Finally, as the transition functions $\gamma_{\alpha\beta}\colon U_\alpha \cap U_\beta \to S^1$ are real (mod 1) valued, one can check that $[\Omega] \in H^2(M^{2n},\mathbb{Z})$ (see e.g. Kobayashi [31] for details).

CHAPTER II

ALMOST CONTACT MANIFOLDS

1. Structural Group of a Contact Manifold.

Before defining the notion of an almost contact structure in section 2, we will show that the structural group of the tangent bundle of a contact manifold can be reduced to $U(n) \times 1$ (Gray [23], Chern [15]).

Theorem. Let M^{2n+1} be a contact manifold. Then the structural group of TM^{2n+1} can be reduced to $U(n) \times 1$.

Proof. Since M^{2n+1} is orientable we may reduce the structural group of TM^{2n+1} from $G\ell(2n+1,\mathbb{R})$ to $SO(2n+1,\mathbb{R})$. Moreover, as we have seen, the contact form η determines complementary distributions of dimension $2n$ and 1 respectively. Thus, we may reduce the structural group to $SO(2n,\mathbb{R}) \times SO(1,\mathbb{R}) = SO(2n,\mathbb{R}) \times 1$. Now let $\{U_\alpha\}$ be an open covering of M^{2n+1} such that $d\eta$ is given on U^α by $\sum_{i=1}^n \theta^i \wedge \theta^{n+i}$ where the θ^i's and θ^{n+i}'s are 1-forms on U_α and let $G_{\alpha\beta}\colon U_\alpha \cap U_\beta \to SO(2n,\mathbb{R}) \cong SO(2n,\mathbb{R}) \times 1$ be the transition functions for TM^{2n+1} with respect to the covering $\{U_\alpha\}$ where for convenience we will regard $G_{\alpha\beta}$ as a matrix in $SO(2n,\mathbb{R})$. If now F is the matrix of components of $\sum_{i=1}^n \theta^i \wedge \theta^{n+i}$ (restricted to D), then $G_{\alpha\beta}F = F\,G_{\alpha\beta}$; but

$F = \frac{1}{2}\begin{pmatrix} O & I \\ -I & O \end{pmatrix}$ where I is the $n \times n$ identity matrix and therefore $G_{\alpha\beta} = \begin{pmatrix} A & B \\ -B & A \end{pmatrix}$ where $A = (a_{ij})$ and $B = (b_{ij})$ are $n \times n$ matrices. Now set $\psi(G_{\alpha\beta}) = (a_{ij} + \sqrt{-1}\, b_{ij})$. Then $\overline{\psi(G_{\alpha\beta})}^{\,t} = \psi(G_{\alpha\beta}{}^{t}) = \psi(G_{\alpha\beta}{}^{-1}) = \psi(G_{\alpha\beta})^{-1}$, that is $\psi(G_{\alpha\beta}) \in U(n)$ and ψ^{-1} maps $U(n)$ onto the set of all matrices $G \in SO(2n,\mathbb{R})$ such that $GF = FG$. Thus $\psi\colon \{G \in SO(2n,\mathbb{R}) \mid GF = FG\} \to U(n)$ is an isomorphism onto $U(n)$ and so the structural group of TM^{2n+1} is reducible to $U(n) \times 1$.

2. Almost Contact Structures

In view of the theorem just proved in section 1 we are led to the notion of an almost contact structure. A differentiable manifold M^{2n+1} is said to have an almost contact structure if the structural group of its tangent bundle is reducible to $U(n) \times 1$ (Gray [23]).

In this and the next section we give alternate definitions of the notion of an almost contact structure which are particularly well suited for much of our work. Therefore let us provisionally define the notion of a (φ, ξ, η)-structure. A differentiable manifold M^{2n+1} is said to have a (φ, ξ, η)-structure if it admits a field φ of endomorphisms of the tangent spaces, a vector field ξ, and a 1-form η satisfying

$$\eta(\xi) = 1 \tag{1}$$

and

$$\varphi^2 = -I + \eta \otimes \xi \tag{2}$$

where I denotes the identity transformation (Sasaki [56]). It seems customary to include also

$$\varphi\xi = 0 \tag{3}$$

and

$$\eta \circ \varphi = 0 \tag{4}$$

in the definition of a (φ,ξ,η)-structure, but these are deducible from (1) and (2) as we show in the following proposition.

<u>Proposition</u>. Suppose M^{2n+1} has a (φ,ξ,η)-structure. Then $\varphi\xi = 0$ and $\eta \circ \varphi = 0$. Moreover the endomorphism φ has rank $2n$.

<u>Proof</u>. First note that (1) and (2) give $\varphi^2\xi = -\xi + \eta(\xi)\xi = 0$ and hence either $\varphi\xi = 0$ or $\varphi\xi$ is a non-trivial eigenvector of φ corresponding to eigenvalue 0. So by (2) again $0 = \varphi^2\varphi\xi = -\varphi\xi + \eta(\varphi\xi)\xi$ or $\varphi\xi = \eta(\varphi\xi)\xi$. Now if $\varphi\xi$ is a non-trivial eigenvector of the eigenvalue 0, $\eta(\varphi\xi) \neq 0$ and therefore $0 = \varphi^2\xi = \eta(\varphi\xi)\varphi\xi = (\eta(\varphi\xi))^2\xi \neq 0$, a contradiction. Thus, $\varphi\xi = 0$.

Now since $\varphi\xi = 0$, we also have from (2) that $\eta(\varphi X)\xi = \varphi^3X + \varphi X = -\varphi X + \varphi(\eta(X)\xi) + \varphi X = 0$ for any vector field X and hence $\eta \circ \varphi = 0$.

Finally since $\varphi\xi = 0$, $\xi \neq 0$ everywhere, rank $\varphi < 2n + 1$. If a vector field $\bar{\xi}$ satisfies $\varphi\bar{\xi} = 0$, (2) gives $0 = -\bar{\xi} + \eta(\bar{\xi})\xi$, that is $\bar{\xi}$ is proportional to ξ and so $\mathrm{rank}(\varphi) = 2n$, completing the proof.

If a manifold M^{2n+1} with a (φ,ξ,η)-structure admits a Riemannian metric g such that

$$g(\varphi X,\varphi Y) = g(X,Y) - \eta(X)\eta(Y) \tag{5}$$

for any vector fields X,Y, then M^{2n+1} is said to have a

(φ,ξ,η,g)-<u>structure or an almost contact metric structure</u> and g is called a <u>compatible metric</u> (Sasaki [56]). Setting $Y = \xi$, an immediate consequence is that η is the covariant form of ξ, that is

$$\eta(X) = g(\xi,X). \tag{6}$$

We now show that such a metric always exists on a manifold with a (φ,ξ,η)-structure.

<u>Proposition</u>. If M^{2n+1} is a manifold with a (φ,ξ,η)-structure, then M^{2n+1} admits a Riemannian metric g such that $g(\varphi X,\varphi Y) = g(X,Y) - \eta(X)\eta(Y)$.

<u>Proof</u>. Let h' be any Riemannian metric on M^{2n+1} and define h by $h(X,Y) = h'(\varphi^2 X,\varphi^2 Y) + \eta(X)\eta(Y)$. Then $h(\xi,X) = \eta(X)$ and it is easy to check that h is a Riemannian metric. Now define g by

$$g(X,Y) = \frac{1}{2}(h(X,Y) + h(\varphi X,\varphi Y) + \eta(X)\eta(Y)).$$

Again g is clearly a Riemannian metric and

$$\begin{aligned} g(\varphi X,\varphi Y) &= \frac{1}{2}(h(\varphi X,\varphi Y) + h(-X+\eta(X)\xi,-Y+\eta(Y)\xi)) \\ &= \frac{1}{2}(h(\varphi X,\varphi Y) + h(X,Y) - 2\eta(X)\eta(Y) + \eta(X)\eta(Y)) \\ &= g(X,Y) - \eta(X)\eta(Y) \end{aligned}$$

as desired.

One should note that the above metric is, of course, not unique.

On a manifold M^{2n+1} with a (φ,ξ,η,g)-structure we can find a particularly useful local orthonormal basis. Let U_α be a coordinate neighborhood and take X_1 a unit vector field on U_α orthogonal to ξ. Then by (5) and (6) φX_1 is also a unit vector field orthogonal to both ξ and X_1. Now take X_2 to be a unit vector field on U_α orthogonal to ξ, X_1 and φX_1, then φX_2 is a unit vector field orthogonal to $\xi, X_1, \varphi X_1$ and X_2. Proceeding in this way we obtain a local orthonormal basis $\{X_i, X_{i^*} = \varphi X_i, \xi\}$, $i = 1, \cdots, n$ called a φ-__basis__.

We now show that the notions of an almost contact structure and a (φ,ξ,η)-structure are equivalent (Sasaki [56]).

__Theorem__. If a manifold M^{2n+1} has a (φ,ξ,η)-structure, then the structural group of its tangent bundle is reducible to $U(n) \times 1$. Conversely an almost contact manifold carries a (φ,ξ,η)-structure.

__Proof__. Suppose M^{2n+1} has a (φ,ξ,η)-structure and let g be a compatible metric. Let $\{U_\alpha\}$ be an open covering of M^{2n+1} by neighborhoods with φ-bases $\{X_i, X_{i^*}, \xi\}$ and $\{\bar{X}_i, \bar{X}_{i^*}, \xi\}$ on U_α and U_β respectively. With respect to these bases the matrix of φ is

$$\begin{pmatrix} 0 & -I & 0 \\ I & 0 & 0 \\ 0 & 0 & 0 \end{pmatrix} . \tag{7}$$

If X is any vector in $T_m M^{2n+1}$, $m \in U_\alpha \cap U_\beta \neq \emptyset$ and (X) and $(\bar{X})$ denote its column vectors of components with respect to the φ-bases on U_α and U_β respectively, then

$$(\bar{X}) = \begin{pmatrix} A & B & 0 \\ C & D & 0 \\ 0 & 0 & 1 \end{pmatrix} (X)$$

where A,B,C,D are $n \times n$ matrices, and the

$(2n+1) \times (2n+1)$ matrix $\begin{pmatrix} A & B & 0 \\ C & D & 0 \\ 0 & 0 & 1 \end{pmatrix}$ is orthogonal.

Clearly this matrix must commute with (7) and hence, as in section 1, $D = A$ and $C = -B$.

Therefore $\begin{pmatrix} A & B & 0 \\ C & D & 0 \\ 0 & 0 & 1 \end{pmatrix}$ belongs to $U(n) \times 1$.

Conversely suppose M^{2n+1} has an almost contact structure and that $\{U_\alpha\}$ is an open covering of M^{2n+1} such that we can choose local orthonormal bases which transform in the overlaps of the neighborhoods by the action of

$U(n) \times 1$. Then with respect to such bases we can define a field of endorphisms φ_α over U_α by the matrix

$$\begin{pmatrix} 0 & -I & 0 \\ I & 0 & 0 \\ 0 & 0 & 0 \end{pmatrix},$$

but this matrix commutes with $U(n) \times 1$ and hence the φ_α's determine a global field of endomorphisms φ. Now define ξ and η by their matrices of components over U_α, namely

$\begin{pmatrix} 0 \\ \vdots \\ 0 \\ 1 \end{pmatrix}$ and $(0,\cdots,0,1)$ respectively. Again ξ and η are globally defined and clearly $\varphi^2 = -I + \eta \otimes \xi$ and $\eta(\xi) = 1$.

Thus the notions of an almost contact structure and a (φ,ξ,η)-structure are equivalent and we will often refer to an almost contact structure (φ,ξ,η).

3. Contact Metric Structures

One sometimes finds in the literature a third definition of an almost contact structure. In particular, a differentiable manifold M^{2n+1} is said to have an almost contact structure if it admits a global 1-form η and a global 2-form Φ such that $\eta\wedge\Phi^n \neq 0$ everywhere (see e.g. Takizawa [69]; Libermann [36] calls this an almost cosymplectic structure).

Suppose M^{2n+1} has an almost contact structure (φ,ξ,η) and let g be a compatible metric. Then we define a 2-form Φ on M^{2n+1} by

$$\Phi(X,Y) = g(X,\varphi Y).$$

The skew-symmetry of Φ is immediate from equations (5),(4) and (2), for $g(X,\varphi Y) = g(\varphi X,\varphi^2 Y) = -g(\varphi X, Y)$. We call Φ the <u>fundamental 2-form</u> of the almost contact metric structure (φ,ξ,η,g). Clearly as φ has rank $2n$, we have $\eta\wedge\Phi^n \neq 0$; thus, a manifold M^{2n+1} with an almost contact structure (φ,ξ,η) is almost contact in the present sense. In this section we shall prove the converse; however, our major purpose is to show the existence of an almost contact metric structure (φ,ξ,η,g) on a contact manifold with contact form η such that $\Phi = d\eta$.

<u>Proposition</u>. Let M^{2n+1} be a differentiable manifold admitting a global 1-form η and a global 2-form Φ such

that $\eta\wedge\Phi^n \neq 0$ everywhere. Then M^{2n+1} admits an almost contact structure. If M^{2n+1} is a contact manifold with contact form η, then there exists an almost contact metric structure (φ,ξ,η,g) (same η) such that the fundamental 2-form $\Phi = d\eta$.

Proof. In the first case, since M^{2n+1} is orientable and Φ is a 2-form of maximal rank, there exists a non-vanishing vector field ξ' such that $\Phi(\xi',X) = 0$ for all vector fields X. Let h be any Riemannian metric on M^{2n+1}, normalize ξ' to a unit vector field ξ and define a 1-form η' by $\eta'(X) = h(X,\xi)$. In the contact case we already have the characteristic vector field ξ. Thus, if h' is any Riemannian metric on M^{2n+1}, h defined by

$$h(X,Y) = h'(-X+\eta(X)\xi,\ -Y+\eta(Y)\xi) + \eta(X)\eta(Y)$$

is a Riemannian metric such that $\eta(X) = h(X,\xi)$.

Thus, in the first case taking the given form Φ and in the second case setting $\Phi = d\eta$, Φ is a symplectic form on the orthogonal complement of ξ and hence there exists a metric g' and an endomorphism φ on the orthogonal complement of ξ such that $g'(X,\varphi Y) = \Phi(X,Y)$ and $\varphi^2 = -I$. Extending g' to a metric g agreeing with h in the direction ξ and extending φ so that $\varphi\xi = 0$, we have that (φ,ξ,η') is an almost contact structure in the first case and (φ,ξ,η,g) an almost contact metric structure with $\Phi = d\eta$ in the contact case.

An almost contact metric structure with $\Phi = d\eta$ is called an <u>associated almost contact metric structure</u> for the contact structure η or more simply we will refer to a <u>contact metric structure</u> (φ, ξ, η, g) or an <u>associated metric</u> g.

The associated metric is, of course, not unique as we could have started with any Riemannian metric h' on M^{2n+1}. The contact structure does, however, impose some restrictions on the behavior of the resulting associated metric g; for example g cannot be flat for contact manifolds of dimension ≥ 5 as we will prove in Chapter VI. In example C below we shall see that there exist almost contact metric structures (φ, ξ, η, g) with $\eta \wedge (d\eta)^n \neq 0$ but $\Phi \neq d\eta$.

In the case of an almost contact structure (φ, ξ, η) we note that η and φ are not independent. In fact for a compatible metric g, we will show that η is $\dfrac{*\Phi^n}{|*\Phi^n|}$ to within sign (e.g. [8]) where $*$ is the Hodge star operator of g. Let $\epsilon(\eta)$ denote the operation of exterior multiplication by η and $\iota(\eta) = (-1)^{(2n+1)(p+1)} * \epsilon(\eta) *$ its dual interior product operator on p-forms. Since $\Phi(\xi, X) = 0$, we have $\iota(\eta)\epsilon(\eta)\Phi^n = \Phi^n$. Therefore

$$*\Phi^n = *\iota(\eta)**\epsilon(\eta)**\Phi^n = \epsilon(\eta)\iota(\eta)*\Phi^n = f\eta$$

where $f = \iota(\eta)*\Phi^n$. Now

$$\langle \Phi^n, \Phi^n \rangle * 1 = \Phi^n \wedge * \Phi^n = f\eta \wedge \Phi^n = (f * \varepsilon(\eta) * * \Phi^n) * 1$$

$$= (f\iota(\eta) * \Phi^n) * 1 = f^2 * 1$$

and hence $f = \pm |\Phi^n|$ giving $\eta = \pm \dfrac{*\Phi^n}{|*\Phi^n|}$. The $\pm$ sign is to be expected as a change of sign in η and ξ in an almost contact metric structure gives another almost contact metric structure.

4. Examples

A. $\mathbb{R}^{2n+1}$. In example I.2.A we considered $\mathbb{R}^{2n+1}$ with its usual contact structure $dz - \sum_{i=1}^{n} y^i dx^i$ and we saw that its contact distribution D is spanned by $\frac{\partial}{\partial x^i} + y^i \frac{\partial}{\partial z}$ and $\frac{\partial}{\partial y^i}$, $i = 1, \cdots, n$. For normalization convenience, we take as the standard contact structure on $\mathbb{R}^{2n+1}$ the 1-form $\eta = \frac{1}{2}(dz - \sum_{i=1}^{n} y^i dx^i)$. The characteristic vector field ξ is then $2\, \partial/\partial z$ and the Riemannian metric

$$g = \frac{1}{4}\Big(\eta \otimes \eta + \sum_{i=1}^{n} ((dx^i)^2 + (dy^i)^2)\Big)$$

gives a contact metric structure on $\mathbb{R}^{2n+1}$. For reference purposes, we give the matrix of components of g, namely

$$\frac{1}{4}\begin{pmatrix} \delta_{ij} + y^i y^j & 0 & -y^i \\ 0 & \delta_{ij} & 0 \\ -y^j & 0 & 1 \end{pmatrix}.$$

The tensor field φ is given by the matrix

$$\begin{pmatrix} 0 & \delta_{ij} & 0 \\ -\delta_{ij} & 0 & 0 \\ 0 & y^j & 0 \end{pmatrix}$$

and the vector fields $X_i = 2\, \partial/\partial y^i$, $X_{n+i} = 2(\partial/\partial x^i + y^i \partial/\partial z)$ $i = 1, \cdots, n$ and ξ form a φ-basis for the contact metric structure.

The Riemannian metric g given here has the following properties. The vector field ξ generates a 1-parameter group of isometries of g, that is ξ is a Killing vector field. The sectional curvature of any plane section containing the vector ξ is equal to 1. The sectional curvature of any plane section spanned by a vector X orthogonal to ξ and φX is equal to -3.

Again the associated metric of a contact structure is not unique and in section 2 of Chapter VI we shall give an associated metric of the contact form η on R^{2n+1} which is less standard but which has some interesting and basic properties.

B. $M^{2n+1} \subset M^{2n+2}$ almost complex, S^{2n+1}. In this example we prove a result of Tashiro [75] that every C^∞ orientable hypersurface of an almost complex manifold has an almost contact structure.

Recall that an almost complex structure on an even-dimensional manifold is a reduction of the structural group to $U(n)$ which is equivalent to a field J of endomorphisms of the tangent spaces such that $J^2 = -I$. Moreover such a manifold admits a metric G, called a Hermitian metric, satisfying $G(X,Y) = G(JX,JY)$ and we say that (J,G) is an almost Hermitian structure on M^{2n}.

Now let M^{2n+2} be an almost complex manifold with almost complex structure J and $\iota : M^{2n+1} \to M^{2n+2}$ a C^∞ orientable hypersurface. We can choose a vector field C

along M^{2n+1} transverse to M^{2n+1} in M^{2n+2} such that the transform JC of C by J is tangent to M^{2n+1}. Indeed, if $J\iota_*X$ were tangent to M^{2n+1} for every vector field X on M^{2n+1}, then $J\iota_*X = \iota_*fX$ for some tensor field f of type (1,1) on M^{2n+1}. Applying J to this equation gives $f^2 = -I$, that is, f is an almost complex structure on M^{2n+1} which is impossible. Thus, there exists a vector field ξ on M^{2n+1} such that $C = J\iota_*\xi$ is transverse to M^{2n+1}. Now define a tensor field φ of type (1,1) and a 1-form η on M^{2n+1} by

$$J\iota_*X = \iota_*\varphi X + \eta(X)C \; ; \tag{8}$$

then applying J we have

$$-\iota_*X = \iota_*\varphi^2X + \eta(\varphi X)C - \eta(X)\iota_*\xi$$

and hence $\varphi^2 = -I + \eta \otimes \xi$ and $\eta \circ \varphi = 0$. Taking $X = \xi$ in equation (8) we have $C = \iota_*\varphi\xi + \eta(\xi)C$, so that $\varphi\xi = 0$ and $\eta(\xi) = 1$. Thus, (φ, ξ, η) defines an almost contact structure on M^{2n+1}.

If we regard M^{2n+2} as an almost Hermitian manifold, take C to be a unit normal to M^{2n+1}. Then JC is tangent to M^{2n+1} so that $JC = -\iota_*\xi$ defines ξ and the rest of the above procedure is repeated. In this case the induced metric $g = \iota^*G$ is compatible with the almost contact structure (φ, ξ, η) since

$$g(X,Y) = G(\iota_*X, \iota_*Y) = G(J\iota_*X, J\iota_*Y)$$
$$= g(\varphi X, \varphi Y) + \eta(X)\eta(Y)$$

We can construct the usual contact structure on an odd-dimensional sphere in this way. Let $S^{2n+1}(r)$ be a sphere of radius r in $\mathbb{R}^{2n+2} \approx \mathbb{C}^{n+1}$ with its usual Kähler structure J, i.e. J is parallel with respect to the Riemannian connection D of the Euclidean metric on $\mathbb{R}^{2n+2}$. Then the structure induced as above with respect to the unit outer normal C is an almost contact metric structure (φ,ξ,η,g) and clearly η is the standard contact form (cf. example I.2.D.). Since S^{2n+1} is umbilical in $\mathbb{R}^{2n+2}$, the second fundamental form h satisfies $h = -\frac{1}{r}g$. Thus, using the fact that J is parallel and the Gauss-Weingarten equations we have

$$0 = (D_{\iota_*X}J)\iota_*\xi = D_{\iota_*X}C - J(\iota_*\nabla_X\xi + h(X,\xi)C)$$
$$= \frac{1}{r}\iota_*X - \iota_*\varphi\nabla_X\xi - \frac{1}{r}\eta(X)\iota_*\xi$$

where ∇ is the Riemannian connection of g. Applying φ we have $\nabla_X\xi = -\frac{1}{r}\varphi X$; since φ has rank $2n$ we again see that $\eta\wedge(d\eta)^n \neq 0$. The almost contact metric structure (φ,ξ,η,g) is not an associated one for $r \neq 1$ as $d\eta = \frac{1}{r}\Phi$ but, of course, the structure $\bar{\eta} = \frac{1}{r}\eta$, $\bar{\xi} = r\xi$, $\bar{\varphi} = \varphi$ and the homothetic change of metric $\bar{g} = \frac{1}{r^2}g$ gives a contact metric structure $(\bar{\varphi},\bar{\xi},\bar{\eta},\bar{g})$. Alternatively the metric

$g' = \frac{1}{r}g + (1-\frac{1}{r})\eta \otimes \eta$ is an associated one for the induced contact form η on $S^{2n+1}(r)$.

C. $S^5 \subset S^6$. We have just seen that an orientable hypersurface of an almost complex manifold carries an almost contact structure and that an odd-dimensional sphere inherits a contact structure in this way from the Kähler structure on $\mathbb{C}^{n+1}$. Now the six-dimensional sphere also carries an almost complex structure, and hence S^5 inherits an almost contact structure from S^6 when considered as a hypersurface [5].

First let us recall the usual almost complex structure on S^6. $\mathbb{R}^7$ considered as the space of imaginary Cayley numbers $\mathbb{O}'$ admits a vector product $\times$ defined by the imaginary part of the product of the two vectors multiplied as Cayley numbers. Letting N denote the unit outer normal to S^6 in $\mathbb{R}^7$ and μ the imbedding $\mu_* JX = N \times \mu_* X$ defines an almost complex structure J on S^6.

Let $(x^1,\cdots,x^7)$ be cartesian coordinates on $\mathbb{R}^7$ and $\Sigma_{i=1}^7 (x^i)^2 = 1$ the equation of S^6. Similarly let $e_1,\cdots, e_7$ be the basis elements of $\mathbb{O}'$. Note that $\mathbb{O}$ has the following representation as $\mathbb{C}^4$

$$x^0 + x^1e_1 + x^2e_2 + x^3e_3 + x^4e_4 + x^5e_5 + x^6e_6 + x^7e_7$$

$$= (x^0 + x^7e_7) + (x^2 + x^5e_7)e_2 + (x^3 + x^4e_7)e_3 + (x^6 + x^1e_7)e_6$$

(a multiplication table for the Cayley numbers may be found

in [30], p. 137). Finally by a <u>Cayley triangle</u> we mean an orthonormal triple (a,b,c) of imaginary Cayley numbers such that c is orthogonal to ab. It is then easy to check that ac and bc are orthogonal (see e.g. [53], p. 282).

Now consider S^5 as a totally geodesic hypersurface of S^6 given by $x^7 = 0$. Taking $C = -\partial/\partial x^7$ as the unit normal, let (φ,ξ,η,g) denote the induced almost contact metric structure on S^5. In particular

$$-\mu_*\iota_*\xi = \mu_* JC = -N \times \partial/\partial x^7 = -\Sigma_{i=1}^{6} x^i \frac{\partial}{\partial x^i} \times \frac{\partial}{\partial x^7}$$

$$= x^6 \frac{\partial}{\partial x^1} - x^5 \frac{\partial}{\partial x^2} - x^4 \frac{\partial}{\partial x^3} + x^3 \frac{\partial}{\partial x^4} + x^2 \frac{\partial}{\partial x^5} - x^1 \frac{\partial}{\partial x^6}$$

from which we see that η is the usual contact form on S^5. Compare this with the usual construction of (φ',ξ,η,g) on $S^5 \subset \mathbb{R}^6(x^7=0) \approx \mathbb{C}^3 = \{x^2+\sqrt{-1}\,x^5,\ x^3+\sqrt{-1}\,x^4,\ x^6+\sqrt{-1}\,x^1\}$; the almost complex structure J' here is given by $J'\frac{\partial}{\partial x^2} = \frac{\partial}{\partial x^5}$, $J'\frac{\partial}{\partial x^3} = \frac{\partial}{\partial x^4}$ and $J'\frac{\partial}{\partial x^6} = \frac{\partial}{\partial x^1}$ (considering $\mathbb{C}^3 \subset \mathbb{C}^4 \approx \mathbb{O}$, J' is just left multiplication by e_7).

Now let X be a vector field tangent to S^5 and orthogonal to ξ and for simplicity identify notationally X, $\iota_* X$ and $\mu_*\iota_* X$. (N,C,X) is then a Cayley triangle in $\mathbb{O}'$ and consequently $\mu_*\iota_*\varphi X = N \times X$ and $\mu_*\iota_*\varphi' X = \mu_* J' X = -C \times X$ are orthogonal (cf. Sato [63]). Thus, S^5 carries two almost contact metric structures, one of which is an associated structure for the usual contact structure and the other is not eventhough the form η in (φ,ξ,η,g) is a contact form. The difference between φ and φ' will be seen again in example IV.6.B by comparison of their covariant derivatives.

D. $M^{2n}\times\mathbb{R}$. Let M^{2n} be an almost complex manifold with almost complex structure J. We consider the manifold $M^{2n+1}=M^{2n}\times\mathbb{R}$, though a similar construction can be carried out for the product $M^{2n}\times S^1$. Denote a vector field on M^{2n+1} by $(X,f\frac{d}{dt})$ where X is tangent to M^{2n}, t the coordinate of $\mathbb{R}$ and f a C^∞ function on M^{2n+1}. Then taking $\eta=dt$, $\xi=(0,\frac{d}{dt})$ and $\varphi(X,f\frac{d}{dt})=(JX,0)$, (φ,ξ,η) is clearly an almost contact structure on M^{2n+1}.

E. Parallelizable manifolds. Let M^{2n+1} be an odd-dimensional parallelizable manifold and denote by $X_1,\cdots,X_{2n+1}$ a set of parallelizing vector fields. Moreover $g(X_A,X_B)=\delta_{AB}$, $A,B=1,\cdots,2n+1$ defines a Riemannian metric on M^{2n+1}. Let $\xi=X_{2n+1}$ and η its covariant form with respect to g. Similarly let ω^i be the covariant form of X_i, $i=1,\cdots,n$ and ω^{i^*} that of $X_{n+i}\equiv X_{i^*}$. Then define φ by $\varphi=\sum_{i=1}^{n}(\omega^i\otimes X_{i^*}-\omega^{i^*}\otimes X_i)$ and it is easy to check that (φ,ξ,η,g) is an almost contact metric structure.

In particular any odd-dimensional Lie group carries an almost contact structure.

CHAPTER III

GEOMETRIC INTERPRETATION OF THE CONTACT CONDITION

1. Integral Submanifolds of the Contact Distribution.

Let M^{2n+1} be a contact manifold with contact form η. We have seen that $\eta = 0$ defines a 2n-dimensional distribution or subbundle D of the tangent bundle called the contact distribution. Roughly speaking the condition $\eta\wedge(d\eta)^n \neq 0$ means that D is as far from being integrable as possible and in this chapter we consider briefly the integral submanifolds of D. In particular we shall see that the maximum dimension of an integral submanifold of D is n (Sasaki [58]).

The phenomena of the non-integrability of D can be easily visualized in R^3 with $\eta = \cos x^3 dx^1 + \sin x^3 dx^2$ as in example I.2.C. In this case D is spanned by $\sin x^3 \partial/\partial x^1 - \cos x^3 \partial/\partial x^2$ and $\partial/\partial x^3$; thus, we see that the planes of D rotate as we move along the x^3-axis.

<u>Theorem</u>. Let M^{2n+1} be a contact manifold with contact form η. Then there exist integral submanifolds of the contact distribution D of dimension n but of no higher dimension.

<u>Proof</u>. Since $\eta\wedge(d\eta)^n \neq 0$ on M^{2n+1}, we can choose local coordinates (x^i, y^i, z), $i = 1,\cdots, n$, such that $\eta = dz - \sum_{i=1}^n y^i dx^i$ on the coordinate neighborhood. Then for a point m with coordinates (x_0^i, y_0^i, z_0) in the

coordinate neighborhood, $x^i = x^i_0, z = z_0$ defines an n-dimensional integral submanifold of D in the neighborhood and a maximal integral submanifold containing this coordinate slice is an integral submanifold of D in M^{2n+1}.

Now suppose M^r is an r-dimensional integral submanifold of D and suppose that $r > n$. Let X_i, $i = 1, \cdots, r$ be r linearly independent local vector fields tangent to M^r and extend these to a basis by $X_{r+1}, \cdots, X_{2n}, X_{2n+1} = \xi$. Then for $i,j = 1,\cdots,r, \eta(X_i) = 0$ and

$$d\eta(X_i,X_j) = \frac{1}{2}(X_i\eta(X_j) - X_j\eta(X_i) - \eta([X_i,X_j])) = 0.$$

Thus, since $r > n, (\eta\wedge(d\eta)^n)(X_1,\cdots,X_{2n+1}) = 0$, a contradiction.

We have just seen that if X,Y are vector fields tangent to an integral submanifold of D, then $\eta(X) = \eta(Y) = 0$ and $d\eta(X,Y) = 0$. These conditions are also sufficient for a submanifold to be an integral submanifold of D, which we state as a proposition.

Proposition. Let $\iota: M^r \to M^{2n+1}$ be an immersed submanifold. Then M^r is an integral submanifold of D if and only if η and $d\eta$ restricted to M^r vanish. Let (φ,ξ,η,g) be an associated almost contact metric structure; then M^r is an integral submanifold of D if and only if every tangent vector X belongs to D and φX is normal to M^r in M^{2n+1}.

Proof. The second statement follows immediately from the first since $d\eta(X,Y) = g(X,\varphi Y)$. The necessity of the first statement has already been noted. The sufficiency is also easy, for if X and Y are vector fields tangent to M^r and η and $d\eta$ vanish on M^r, then

$$0 = d\eta(X,Y) = \frac{1}{2}(X\eta(Y) - Y\eta(X) - \eta([X,Y])) = -\frac{1}{2}\eta([X,Y])$$

so that $[X,Y]$ belongs to D.

Similarly the conditions of η and $d\eta$ vanishing on r linearly independent vectors $(r \leq n)$ at a point is sufficient for the existence of an r-dimensional integral submanifold tangent to them (Sasaki [58]).

Proposition. Let (x^i,y^i,z), $i = 1,\cdots,n$, be local coordinates about $m = (x_0^i,y_0^i,z_0)$ such that $\eta = dz - \sum_{i=1}^n y^i dx^i$ on the coordinate neighborhood. In order that r linearly independent vectors $X_\lambda, \lambda = 1,\cdots,r \leq n$ at m with components $(a_\lambda^i,b_\lambda^i,c_\lambda)$ be tangent to an r-dimensional integral submanifold it is necessary and sufficient that $\eta(X_\lambda) = 0$ and $d\eta(X_\lambda,X_\mu) = 0$, that is

$$c_\lambda = \sum_i y_0^i a_\lambda^i ,$$

$$\sum_i a_\lambda^i b_\mu^i = \sum_i a_\mu^i b_\lambda^i .$$

Proof. Again the necessity is clear. To prove the sufficiency, set $c_{\lambda\mu} = \sum_i a_\lambda^i b_\mu^i$ and choose a sufficiently small neighborhood V of the origin of $\mathbb{R}^r$ with coordinates $(u^1,\cdots, u^r)$ such that

$$x^i = x_0^i + \sum_{\lambda=1}^{r} a_\lambda^i u^\lambda$$

$$y^i = y_0^i + \sum_{\lambda=1}^{r} b_\lambda^i u^\lambda$$

$$z = z_0 + \sum_\lambda c_\lambda u^\lambda + \frac{1}{2}\sum c_{\lambda\mu} u^\lambda u^\mu$$

defines a mapping ι of V into M^{2n+1}. Then $\frac{\partial x^i}{\partial u^\lambda} = a_\lambda^i$, $\frac{\partial y^i}{\partial u^\lambda} = b_\lambda^i$ and

$$\frac{\partial z}{\partial u^\lambda} = c_\lambda + \sum_\mu c_{\lambda\mu} u^\mu = \sum_i y_0^i \frac{\partial x^i}{\partial u^\lambda} + \sum_{i,\mu} \frac{\partial x^i}{\partial u^\lambda}\frac{\partial y^i}{\partial u^\mu} u^\mu = \sum_i y^i \frac{\partial x^i}{\partial u^\lambda}$$

and hence the mapping ι defines an integral submanifold of D tangent to $X_1, \cdots, X_r$ at m.

We are now in a position to show the great abundance of integral submanifolds of D; the result is again due to Sasaki [58]. This abundance is one of the difficulties in studying the integral submanifolds of a contact distribution. Another difficulty is that $d\eta$ vanishes along the submanifold and so one does not have an induced structure. As noted above φX is normal to an integral submanifold for X tangent to it, so loosely speaking the geometry is normal to the submanifold. Thus, in Chapter V we study integral submanifolds of only a very restrictive class of contact manifolds, namely the Sasakian space forms.

Theorem. Let X be a vector at $m \in M^{2n+1}$ belonging to D. Then there exists an r-dimensional integral submanifold $M^r (1 \leq r \leq n)$ of D through m such that X is tangent to M^r.

Proof. Let (x^i, y^i, z) be local coordinates as in the previous proposition and let (a_1^i, b_1^i, c_1) be the components of X with respect to these coordinates. Then, since X belongs to D, $c_1 = \Sigma_i y_0^i a_1^i$. If not all the a_1^i's vanish, choose $a_2^i, \cdots, a_r^i$ such that $\text{rank}(a_\lambda^i) = r$ and define $c_2, \cdots, c_r$ by $c_\lambda = \Sigma_i y_0^i a_\lambda^i$, $\lambda = 2, \cdots, r$. Now define $b_2^i, \cdots, b_r^i$ inductively as follows. We are given the b_1^i's and we take $b_{\mu+1}^i (1 \leq \mu \leq r-1)$ as a set of solutions of

$$\Sigma_i a_1^i t^i = \Sigma_i a_{\mu+1}^i b_1^i$$
$$\vdots$$
$$\Sigma_i a_\mu^i t^i = \Sigma_i a_{\mu+1}^i b_\mu^i$$

which exist as $\text{rank}(a_\lambda^i) = r$. Then $\{(a_\lambda^i, b_\lambda^i, c_\lambda)\}$ satisfy the conditions of the previous proposition and hence we have an integral submanifold M^r with X tangent as desired.

If on the other hand all of the a_1^i's vanish, then c_1 also vanishes, so choosing $b_2^i, \cdots, b_r^i$ such that $\text{rank}(b_\lambda^i) = r$ we again have an r-dimensional integral submanifold with X tangent by the previous proposition.

A diffeomorphism $f: M^{2n+1} \to M^{2n+1}$ is a contact transformation if $f^*\eta = \tau\eta$ for some non-vanishing function τ and f is a strict contact transformation if $\tau \equiv 1$. Clearly

$$f^* d\eta = d\tau \wedge \eta + \tau d\eta,$$

so if f is strict $d\eta$ is invariant. Conversely if $d\eta$ is invariant, $d\tau\wedge\eta + (\tau-1)d\eta = 0$ so that $(\tau-1)\eta\wedge d\eta = 0$ and hence $\tau = 1$ and f is strict. The following lemma is also clear.

Lemma. A diffeomorphism f of a contact manifold M^{2n+1} is a contact transformation if and only if f_*X belongs to D for every X in D.

Thus, a contact transformation maps integral curves of D into integral curves of D as remarked at the end of example I.2.F. We can also give a similar characterization of contact transformations in terms of n-dimensional integral submanifolds (Sasaki [58]).

Theorem. A diffeomorphism f of a contact manifold M^{2n+1} is a contact transformation if and only if f maps n-dimensional integral submanifolds of D onto n-dimensional integral submanifolds of D.

Proof. If f is a contact transformation and M^n an integral submanifold, then for X tangent to M^n, f_*X is also in D so that $f(M^n)$ is an integral submanifold. Conversely given any vector X at m belonging to D, we have seen that there exists an integral submanifold M^n through m with X tangent. Now since $f(M^n)$ is also an integral submanifold, f_*X is in D and f is a contact transformation by the lemma.

2. Examples of Integral Submanifolds

Our first three examples will be integral submanifolds of odd-dimensional spheres with some remarks about the special cases.

A. S^{2n+1}. Consider the space $\mathbb{C}^{n+1}$ of n+1 complex variables and let J denote its usual almost complex structure. Let $S^{2n+1} = \{z \in \mathbb{C}^{n+1} \,|\, |z| = 1\}$. Then as we have seen (example II.4.B) we can give S^{2n+1} its usual contact structure as follows. For every $z \in S^{2n+1}$ and $X \in T_z S^{2n+1}$, set $\xi = -Jz$ and φX the tangential part of JX. Let η be the dual 1-form of ξ and g the standard metric on S^{2n+1}. Then (φ,ξ,η,g) is a contact metric structure on S^{2n+1}. Let L be an (n+1)-dimensional linear subspace of $\mathbb{C}^{n+1}$ passing through the origin and such that JL is orthogonal to L. Then $S^{2n+1} \cap L$ satisfies the condition of the first proposition of section 1 and so is an integral submanifold of D for the contact structure η on S^{2n+1}. Clearly $S^{2n+1} \cap L$ is an n-sphere imbedded as a totally geodesic submanifold of S^{2n+1}, [9].

B. S^5. Let $S^5 = \{z \in \mathbb{C}^3 \,|\, |z| = 1\}$ be the 5-dimensional sphere with the contact metric structure described in example A. If we write $z = (z^1,z^2,z^3)$, the equations $|z^1| = |z^2| = |z^3| = 1/\sqrt{3}$ give an imbedding of a 3-dimensional torus T^3 in S^5 which is minimal (see e.g. Chern, doCarmo and Kobayashi [16]). Moreover ξ is tangent to T^3, and for X orthogonal to ξ and tangent to T^3, φX is normal to T^3 in S^5. Viewing T^3 as a cube with opposite faces identified, ξ is just a "diagonally pointing" vector field. Now consider a 2-dimensional torus T^2 imbedded in T^3 by $\sum_{\alpha=1}^{3} \log \sqrt{3}\, z^\alpha = 2k\pi\sqrt{-1}$ where the logarithm is the multi-valued one and k is an integer. Then T^2 is orthogonal to ξ in T^3 and hence an integral submanifold of S^5. Since $\nabla_X \xi = -\varphi X$, as we saw in example II.4.B, T^2 is totally geodesic in T^3 and hence minimal and not totally geodesic in S^5, [9].

Examples A and B show that S^5 contains both S^2 and T^2 as integral submanifolds of its contact distribution. It is well known that for topological reasons (see e.g. [64]) S^5 does not admit a continuous field of 2-planes. Thus, S^5 cannot be foliated by integral surfaces of its contact structure.

C. S^3. Consider S^3 as the unit sphere in $\mathbb{H}$, the space of quaternions. $\mathbb{H}$ admits three almost complex structures J_1, J_2, J_3 which are Hermitian with respect to the usual metric and such that $J_3 = J_1J_2 = -J_2J_1$. Applying $J_a, a = 1,2,3$ to the unit outer normal C we obtain three orthonormal vector fields on S^3, $\xi_a = -J_aC, a = 1,2,3$. We take as the standard contact metric structure on S^3, the structure $(\varphi, \xi_1, \eta, g)$ induced by J_1. The contact distribution D of η is then spanned by ξ_2 and ξ_3. As a contact distribution it is, of course, not integrable and the fact that $[\xi_2, \xi_3] = 2\xi_1$ is well known and easily checked. The integral curves of ξ_2 and ξ_3 are integral submanifolds of D. Thus we can regard S^3 as folliated by integral submanifolds of D by either the integral curves of ξ_2 or ξ_3.

D. $\mathbb{C}^* \times \mathbb{R}$. As in example I.2.G, $\mathbb{C}^*$ is the set of non-zero complex numbers with polar coordinates (r,θ) and we let z denote the coordinate on $\mathbb{R}$. Then $\eta = dz - r^2 d\theta$ is a contact structure with characteristic vector field $\xi = \partial/\partial z$. D is spanned by $\partial/\partial r$ and $V = \frac{1}{r}\frac{\partial}{\partial\theta} + r\frac{\partial}{\partial z}$. Thus, $\mathbb{C}^* \times \mathbb{R}$ is foliated by the integral curves of $\partial/\partial r$ or of V which are just helicies on the cylinders r = constant. The metric whose components with respect to the coordinates (r,θ,z) are given by the matrix

$$\begin{pmatrix} 1 & 0 & 0 \\ 0 & r^2(1+r^2) & -r^2 \\ 0 & -r^2 & 1 \end{pmatrix}$$

is an associated metric for this structure. The matrix of φ is

$$\begin{pmatrix} 0 & -r & 0 \\ 1/r & 0 & 0 \\ r & 0 & 0 \end{pmatrix}.$$

E. T_1^*M, T_1M. In example I.2.E we saw that the bundle of unit cotangent vectors T_1^*M to a Riemannian manifold carries a contact structure induced by the form $\beta = \sum_i p^i dq^i$ on the cotangent bundle T^*M. Here the p^i are coordinates on the fibres over a coordinate neighborhood U on M and $q^i = x^i \circ \pi$ where π is the projection $\pi: T^*M \to M$ and x^i the local coordinates on U. Clearly β and $d\beta$ vanish on vertical vectors. Thus, the fibres of the unit cotangent bundle are integral submanifolds of the contact distribution on T_1^*M. The same remark clearly holds for the tangent sphere bundle T_1M as well.

F. M^{2n+1} <u>over</u> M^{2n} <u>with Lagrangian foliation</u>.

Let M^{2n} be a symplectic manifold with fundamental 2-form Ω and suppose that $[\Omega] \in H^2(M^{2n}, \mathbb{Z})$. Moreover, suppose that M^{2n} admits a Lagrangian foliation given by functions $f_1, \cdots, f_n$ in involution with the df_i's linearly independent, for example the tori T^2 or T^4 as compact symplectic manifolds with integral class. Recall that two functions f and g are in involution if their Poisson bracket $\{f,g\} = \Omega(X_f, X_g)$ vanishes where the vector field X_f is defined by $\Omega(X_f, X) = Xf$ for all vector fields X.

Let M^{2n+1} be the corresponding principal circle bundle and η the contact form as in example I.2.H. Let $X_i = \tilde{\pi} X_{f_i}$ where $\tilde{\pi}$ denotes the horizontal lift with respect to the connection form η. Then $\eta(X_i) = 0$ and $d\eta(X_i,X_j) = \Omega(X_{f_i},X_{f_j}) \circ \pi = 0$ and hence $\eta([X_i,X_j]) = 0$ and $[X_i,X_j] = \tilde{\pi}[X_{f_i},X_{f_j}]$. Thus, the Lagrangian foliation on M^{2n} lifts to a foliation by integral submanifolds of η

CHAPTER IV

K-CONTACT AND SASAKIAN STRUCTURES

1. Normal Almost Contact Structures.

In Chapter II we studied almost contact manifolds, that is manifolds with structural group $U(n) \times 1$ and hence these manifolds can be thought of as odd-dimensional analogues of almost complex manifolds. We now consider those almost contact manifolds which are, in the sense to be defined, analogous to the complex manifolds.

As is well known an almost complex structure need not come from a complex structure, i.e. a differentiable structure on a manifold M^{2n} modeled on the pseudogroup of holomorphic maps of open sets of $\mathbb{C}^n$ onto open sets of $\mathbb{C}^n$. A celebrated theorem of Newlander and Nirenberg [44] states that an almost complex structure J of class $C^{2n+\alpha}$ with vanishing Nijenhuis torsion is integrable, i.e. is the associated almost complex structure of a complex structure. The Nijenhuis torsion $[h,h]$ of a tensor field h of type (1,1) is a tensor field of type (1,2) given by

$$[h,h](X,Y) = h^2[X,Y] + [hX,hY] - h[hX,Y] - h[X,hY].$$

All manifolds under our consideration are of class C^∞, so the Newlander-Nirenberg theorem applies. For detailed studies of complex manifolds see, for example [19],[33] or [83].

Let M^{2n+1} be an almost contact manifold with almost contact structure (φ, ξ, η) and consider the manifold $M^{2n+1} \times \mathbb{R}$. We denote a vector field on $M^{2n+1} \times \mathbb{R}$ by $(X, f\frac{d}{dt})$ where X is tangent to M^{2n+1}, t the coordinate of $\mathbb{R}$ and f a C^∞ function on $M^{2n+1} \times \mathbb{R}$. Define an almost complex structure J on $M^{2n+1} \times \mathbb{R}$ by

$$J(X, f\frac{d}{dt}) = (\varphi X - f\xi, \eta(X)\frac{d}{dt}),$$

that $J^2 = -I$ is easy to check. If now J is integrable, we say that the almost contact structure (φ, ξ, η) is <u>normal</u> [60].

As the vanishing of the Nijenhuis torsion of J is a necessary and sufficient condition for integrability, we seek to express the condition of normality in terms of the Nijenhuis torsion of φ. Since $[J,J]$ is a tensor field of type (1,2) it suffices to compute $[J,J]((X,0),(Y,0))$ and $[J,J]((X,0),(0,\frac{d}{dt}))$ for vector fields X and Y on M^{2n+1} [60].

$[J,J]((X,0),(Y,0))$

$$= -[(X,0),(Y,0)] + [(\varphi X, \eta(X)\frac{d}{dt}), (\varphi Y, \eta(Y)\frac{d}{dt})]$$

$$- J[(\varphi X, \eta(X)\frac{d}{dt}), (Y,0)] - J[(X,0), (\varphi Y, \eta(Y)\frac{d}{dt})]$$

$$= -([X,Y],0) + ([\varphi X, \varphi Y], (\varphi X\eta(Y) - \varphi Y\eta(X))\frac{d}{dt})$$

$$- (\varphi[\varphi X, Y] + (Y\eta(X))\xi, \eta([\varphi X, Y])\frac{d}{dt})$$

$$- (\varphi[X, \varphi Y] - (X\eta(Y))\xi, \eta([X, \varphi Y])\frac{d}{dt})$$

$$= ([\varphi,\varphi](X,Y) + 2d\eta(X,Y)\xi,\ ((\mathfrak{L}_{\varphi X}\eta)(Y) - (\mathfrak{L}_{\varphi Y}\eta)(X))\frac{d}{dt}) .$$

$$[J,J]((X,0),(0,\frac{d}{dt}))$$

$$= -[(X,0),(0,\frac{d}{dt})] + [(\varphi X,\eta(X)\frac{d}{dt}),(-\xi,0)]$$

$$- J[(\varphi X,\eta(X)\frac{d}{dt}),(0,\frac{d}{dt})] - J[(X,0),(-\xi,0)]$$

$$= (-[\varphi X,\xi],(\xi\eta(X))\frac{d}{dt}) + (\varphi[X,\xi],\eta([X,\xi])\frac{d}{dt})$$

$$= ((\mathfrak{L}_\xi\varphi)X,(\mathfrak{L}_\xi\eta)(X)\frac{d}{dt}) .$$

We are thus lead to define four tensors $N^{(1)}, N^{(2)}, N^{(3)}$ and $N^{(4)}$ by

$$N^{(1)}(X,Y) = [\varphi,\varphi](X,Y) + 2d\eta(X,Y)\xi ,$$

$$N^{(2)}(X,Y) = (\mathfrak{L}_{\varphi X}\eta)(Y) - (\mathfrak{L}_{\varphi Y}\eta)(X) ,$$

$$N^{(3)}(X) = (\mathfrak{L}_\xi\varphi)X ,$$

$$N^{(4)}(X) = (\mathfrak{L}_\xi\eta)(X) .$$

It is clear that the almost contact structure (φ,ξ,η) is normal if and only if these four tensors vanish. We will show however that the vanishing of $N^{(1)}$ implies the vanishing of $N^{(2)}, N^{(3)}$ and $N^{(4)}$, so that the normality condition is

$$[\varphi,\varphi] + 2d\eta \otimes \xi = 0 .$$

The remainder of this section is devoted to proving this and other important properties of the tensors $N^{(1)}$, $N^{(2)}$, $N^{(3)}$ and $N^{(4)}$ [60], [61].

Proposition: For an almost contact structure (φ, ξ, η) the vanishing of $N^{(1)}$ implies the vanishing of $N^{(2)}$, $N^{(3)}$ and $N^{(4)}$.

Proof. Since $N^{(1)} = 0$, we have

$$\begin{aligned} 0 &= [\varphi,\varphi](X,\xi) + 2\,d\eta\,(X,\xi)\,\xi \\ &= -[X,\xi] + \eta([X,\xi])\xi - \varphi[\varphi X,\xi] \\ &\quad + (X\eta(\xi))\xi - (\xi\eta(X))\xi - \eta([X,\xi])\xi \\ &= [\xi,X] + \varphi[\xi,\varphi X] - (\xi\eta(X))\xi \end{aligned} \tag{1}$$

Applying η to this equation we obtain

$$0 = \eta([\xi,X]) - \xi\eta(X)$$

which is just $N^{(4)} = \mathfrak{L}_\xi\eta = 0$. Note also at this point that if we replace X by φX we have

$$\eta([\xi,\varphi X]) = 0. \tag{2}$$

Now applying φ to (1) we have

$$0 = \varphi\mathfrak{L}_\xi X - \mathfrak{L}_\xi\varphi X + \eta([\xi,\varphi X]) = (\mathfrak{L}_\xi\varphi)X$$

and hence $N^{(3)} = 0$. Finally using $N^{(1)} = 0$ again,

$$\begin{aligned} 0 &= [\varphi,\varphi](\varphi X,Y) + 2\,d\eta\,(\varphi X,Y)\,\xi \\ &= -[\varphi X,Y] + \eta([\varphi X,Y])\xi + [-X+\eta(X)\xi,\varphi Y] \\ &\quad - \varphi[-X+\eta(X)\xi,Y] - \varphi[\varphi X,\varphi Y] \end{aligned}$$

$$+ (\varphi X\eta(Y))\xi - (Y\eta(\varphi X))\xi - \eta([\varphi X,Y])\xi$$

$$= - [\varphi X,Y] - [X,\varphi Y] - (\varphi Y\eta(X))\xi - \eta(X)[\varphi Y,\xi]$$

$$- \varphi[-X+\eta(X)\xi,Y] - \varphi[\varphi X,\varphi Y] + (\varphi X\eta(Y))\xi .$$

Applying η to this and using (2) we have

$$\varphi X\eta(Y) - \eta([\varphi X,Y]) - \varphi Y\eta(X) + \eta([\varphi Y,X]) = 0 \qquad (3)$$

giving $N^{(2)} = 0$.

Note that (3) can be written

$$d\eta(\varphi X,Y) + d\eta(X,\varphi Y) = 0 \qquad (4)$$

from which we see that $N^{(2)} = 0$ implies $d\eta(X,\xi) = 0$ and hence (4) gives

$$d\eta(\varphi X,\varphi Y) = d\eta(X,Y) . \qquad (5)$$

Conversely it is easy to check that (5) implies (4). Thus the meaning of the vanishing of $N^{(2)}$ for an almost contact structure (φ,ξ,η) is that $d\eta$ is invariant under φ.

We now consider the case of a contact manifold with contact form η and associated almost contact metric structure (φ,ξ,η,g).

<u>Proposition</u>. Let (φ,ξ,η,g) be a contact metric structure. Then the tensors $N^{(2)}$ and $N^{(4)}$ vanish. Moreover $N^{(3)}$ vanishes if and only if the characteristic vector field ξ is Killing with respect to g.

Proof. In view of the above discussion, to show that vanishing of $N^{(2)}$ it suffices to show that equation (5) holds. We have

$$d\eta(\varphi X, \varphi Y) = \Phi(\varphi X, \varphi Y) = g(\varphi X, \varphi^2 Y)$$
$$= - g(X, \varphi^3 Y) = g(X, \varphi Y) = d\eta(X,Y).$$

In section I.1 we saw that $N^{(4)} = \mathfrak{L}_\xi \eta = 0$ for any contact structure.

Now since $\eta^{(4)} = 0$ we automatically have

$$(\mathfrak{L}_\xi g)(X,\xi) = \xi\eta(X) - \eta([\xi,X]) = (\mathfrak{L}_\xi \eta)(X) = 0.$$

We saw in section I.1 that $d\eta$ is invariant under the 1-parameter group of ξ and hence

$$0 = (\mathfrak{L}_\xi \Phi)(X,Y) = \xi g(X,\varphi Y) - g([\xi,X],\varphi Y) - g(X,\varphi[\xi,Y])$$
$$= (\mathfrak{L}_\xi g)(X,\varphi Y) + g(X,(\mathfrak{L}_\xi \varphi)Y).$$

Thus $N^{(3)} = \mathfrak{L}_\xi \varphi = 0$ if and only if ξ is a Killing vector field.

Next we will establish a formula for the covariant derivative of φ for a general almost contact metric structure (φ,ξ,η,g).

Lemma. For an almost contact metric structure (φ,ξ,η,g) the covariant derivative of φ is given by

$$2g((\nabla_X\varphi)Y,Z) = 3\,d\Phi(X,\varphi Y,\varphi Z) - 3\,d\Phi(X,Y,Z)$$
$$+ g(N^{(1)}(Y,Z),\varphi X) + N^{(2)}(Y,Z)\eta(X)$$
$$+ 2\,d\eta(\varphi Y,X)\eta(Z) - 2\,d\eta(\varphi Z,X)\eta(Y). \quad (6)$$

Proof. Recall that the Riemannian connection ∇ of g is given by

$$2g(\nabla_X Y,Z) = Xg(Y,Z) + Yg(X,Z) - Zg(X,Y)$$
$$+ g([X,Y],Z) + g([Z,X],Y) - g([Y,Z],X)$$

and that the coboundary formula for d on a 2-form Φ is

$$d\Phi(X,Y,Z) = \frac{1}{3}\{X\Phi(Y,Z) + Y\Phi(Z,X) + Z\Phi(X,Y)$$
$$- \Phi([X,Y],Z) - \Phi([Z,X],Y) - \Phi([Y,Z],X)\}.$$

Therefore

$$2g((\nabla_X\varphi)Y,Z) = 2g(\nabla_X\varphi Y,Z) + 2g(\nabla_X Y,\varphi Z)$$
$$= Xg(\varphi Y,Z) + \varphi Yg(X,Z) - Zg(X,\varphi Y)$$
$$+ g([X,\varphi Y],Z) + g([Z,X],\varphi Y) - g([\varphi Y,Z],X)$$
$$+ Xg(Y,\varphi Z) + Yg(X,\varphi Z) - \varphi Zg(X,Y)$$
$$+ g([X,Y],\varphi Z) + g([\varphi Z,X],Y) - g([Y,\varphi Z],X)$$
$$= -X\Phi(Y,Z) + \varphi Y(\Phi(\varphi Z,X) + \eta(Z)\eta(X)) - Z\Phi(X,Y)$$
$$- \Phi([X,\varphi Y],\varphi Z) + \eta([X,\varphi Y])\eta(Z)$$
$$+ \Phi([Z,X],Y) - g(\varphi[\varphi Y,Z],\varphi X) + \eta(X)\eta([Z,\varphi Y])$$
$$+ X\Phi(\varphi Y,\varphi Z) - Y\Phi(Z,X) - \varphi Z(\Phi(\varphi Y,X) + \eta(Y)\eta(X))$$

$$+ \Phi([X,Y],Z) - \Phi([\varphi Z,X],\varphi Y) + \eta([\varphi Z,X])\eta(Y)$$

$$- g(\varphi[Y,\varphi Z],\varphi X) + \eta(X)\eta([\varphi Z,Y])$$

$$+ \Phi([Y,Z],X) - g([Y,Z],\varphi X)$$

$$- \Phi([\varphi Y,\varphi Z],X) + g([\varphi Y,\varphi Z],\varphi X)$$

$$+ g(2d\eta(Y,Z)\xi,\varphi X)$$

$$= 3d\Phi(X,\varphi Y,\varphi Z) - 3d\Phi(X,Y,Z) + g(N^{(1)}(Y,Z),\varphi X)$$

$$+ N^{(2)}(Y,Z)\eta(X) + 2d\eta(\varphi Y,X)\eta(Z) - 2d\eta(\varphi Z,X)\eta(Y).$$

In the case of a contact metric structure $\Phi = d\eta$ and $N^{(2)} = 0$, so the above formula becomes

$$2g((\nabla_X\varphi)Y,Z) = g(N^{(1)}(Y,Z),\varphi X) + 2d\eta(\varphi Y,X)\eta(Z) - 2d\eta(\varphi Z,X)\eta(Y). \tag{7}$$

In particular $\nabla_\xi\varphi = 0$. It is also easy to see that on a contact metric manifold the integral curves of ξ are geodesics. Clearly $g(\nabla_\xi\xi,\xi) = 0$ and for X orthogonal to ξ,

$$g(\nabla_\xi\xi,X) = - g(\xi,\nabla_\xi X) = - g(\xi,\nabla_X\xi + [\xi,X])$$

$$= - \eta([\xi,X]) = d\eta(\xi,X) = 0.$$

One final basic property of $N^{(3)} = \mathcal{L}_\xi\varphi$ that we give here is the following [6].

<u>Lemma</u>. On a manifold with a contact metric sgructure $\mathcal{L}_\xi\varphi$ is a symmetric operator.

Proof. We have just noted that for a contact metric structure (φ, ξ, η, g), $\nabla_\xi \varphi = 0$ and $\nabla_\xi \xi = 0$. Thus

$$g((\mathcal{L}_\xi \varphi)X,Y) = g(\nabla_\xi \varphi X - \nabla_{\varphi X}\xi - \varphi\nabla_\xi X + \varphi\nabla_X \xi, Y)$$
$$= g(-\nabla_{\varphi X}\xi + \varphi\nabla_X \xi, Y)$$

which vanishes if either X or Y is ξ. For X and Y orthogonal to ξ, $N^{(2)} = 0$ becomes, by equation (3), $\eta([\varphi X,Y]) + \eta([X,\varphi Y]) = 0$; continuing the computation we then have

$$g((\mathcal{L}_\xi \varphi)X,Y) = \eta(\nabla_{\varphi X}Y) + \eta(\nabla_X \varphi Y)$$
$$= \eta(\nabla_Y \varphi X) + \eta(\nabla_{\varphi Y}X)$$
$$= g((\mathcal{L}_\xi \varphi)Y,X) .$$

In the next section we give an interpretation of normality as an integrability. One might also ask about the integrability of an almost contact structure as a G-structure; that is when do there exist local coordinates $(x^1, \cdots, x^{2n+1})$ such that φ is given by the matrix of components

$$\begin{pmatrix} 0 & I & 0 \\ -I & 0 & 0 \\ 0 & 0 & 0 \end{pmatrix}$$

with respect to the coordinate basis. The condition for this is of course the vanishing of the Nijenhuis torsion of φ (see e.g. [84]). This condition however leads us away from the contact case. In particular, since $\eta \circ \varphi = 0$, we see that locally $\eta = f\,dx^{2n+1}$ for some function f and hence $\eta \wedge d\eta \equiv 0$. Thus the vanishing of $[\varphi, \varphi]$ implies that the 2n-dimension distribution defined by $\eta = 0$ is integrable.

2. Geometric Interpretation of Normality

In the first part of this section we summarize a paper by Andreotti and Hill [3] on complex characteristic coordinates and C - R manifolds. We then prove a theorem of Ianus [29] which states that a normal almost contact manifold is a C - R manifold.

Let U be an open set in $\mathbb{R}^m$ and consider ℓ linearly independent complex vector fields

$$P_k = \sum_{j=1}^{m} c_k^j(x) \frac{\partial}{\partial x^j}, \quad 1 \leq k \leq \ell$$

on U, where the c_k^j's are smooth complex-valued functions on U. A complex characteristic coordinate is a complex-valued C^1 function ζ on U such that $d\zeta \neq 0$ on U and ζ is a solution of the system $P_k\zeta = 0$, $1 \leq k \leq \ell$. Complex characteristic coordinates $\zeta^1, \cdots, \zeta^N$ are said to be functionally independent at x if $d\zeta^1, \cdots, d\zeta^N$ are linearly independent at x. Clearly the maximum number of functionally independent complex characteristic coordinates on U is $N = m - \ell$.

Andreotti and Hill show that if we are given complex vector fields P_k which form an involutive system and if the real and imaginary parts of the coefficients $c_k^j(x)$ are real analytic (slightly less differentiability suffices [3]), then there exist $N = m - \ell$ functionally independent complex characteristic coordinates on U. Thus we shall assume that the P_k's form an involutive system.

Let $\overline{P}_k$ denote the conjugate vector field $\sum_{j=1}^{m} \overline{C}_k^j(x) \frac{\partial}{\partial x^j}$ and assume that the vectors $P_k, \overline{P}_k$, $1 \leq k \leq \ell$ generate a space of constant dimension $\ell + r$ on U.

We denote the matrix of C_k^j's by C and that of the $\overline{C}_k^j$'s by $\overline{C}$. In particular, $\operatorname{rank}\binom{C}{\overline{C}} = \ell + r$.

Now consider the map $\zeta\colon U \to \mathbb{C}^N$ defined by $\zeta(x) = (\zeta^1(x), \cdots, \zeta^N(x))$. We shall show that under our assumptions ζ has rank $N + r$. Let $L = \left(\frac{\partial \zeta^i}{\partial x^j}\right)$, then the columns of L form a basis for the null space Γ_x of C at $x \in U$. Thus the columns of $(L\ \overline{L})$ span $\Gamma_x + \overline{\Gamma}_x$, so that rank $\zeta = \operatorname{rank}(L\overline{L}) = \dim_{\mathbb{C}}(\Gamma_x + \overline{\Gamma}_x)$. Now $\dim_{\mathbb{C}}\Gamma_x = \dim_{\mathbb{C}}\overline{\Gamma}_x = N$ and $\Gamma_x \cap \overline{\Gamma}_x$ is the null space of $\binom{C}{\overline{C}}$ which has dimension $m - (\ell + r) = N - r$. Therefore $\dim_{\mathbb{C}}(\Gamma_x + \overline{\Gamma}_x) = \dim_{\mathbb{C}}\Gamma_x + \dim_{\mathbb{C}}\overline{\Gamma}_x - \dim_{\mathbb{C}}\Gamma_x \cap \overline{\Gamma}_x = N + N - (N - r) = N + r$ as desired.

By shrinking U and relabeling the variables if necessary, we may assume that the first $N + r$ rows of $(L\ \overline{L})$ are linearly independent over $\mathbb{C}$. Set $x' = (x^1, \cdots, x^{N+r})$ and $x'' = (x^{N+r+1}, \cdots, x^m)$. Then the map $\psi\colon U \to \mathbb{C}^N \times \mathbb{R}^{\ell - r}$ defined by $\psi(x', x'') = (\zeta(x', x''), x'')$ is an immersion and if U is sufficiently small an imbedding onto a locally closed (closed subset of some open set) submanifold M^m of $\mathbb{C}^N \times \mathbb{R}^{\ell - r}$. Indeed setting $(L\ \overline{L}) = \begin{pmatrix} L' & \overline{L}' \\ L'' & \overline{L}'' \end{pmatrix}$ corresponding to the splitting of the variables, the Jacobian of ψ is $\begin{pmatrix} L' & \overline{L}' & 0 \\ L'' & \overline{L}'' & I \end{pmatrix}$ which has rank $N+r+(m-(N+r)) = m$. Regarding $\mathbb{R}^{\ell-r}$ as the real part of $\mathbb{C}^{\ell-r}$, ψ is immersion of U into $\mathbb{C}^N \times \mathbb{C}^{\ell-r} = \mathbb{C}^{m-r}$. Also $\overline{M} = \zeta(U)$ for U sufficiently small is a locally closed submanifold of $\mathbb{C}^N$, so that ψ is an isomorphism of U onto an open subset of $\overline{M} \times \mathbb{R}^{\ell-r}$.

Now let M^m be a real C^∞ submanifold imbedded and locally closed in $\mathbb{C}^q$. Locally M^m is given by a system of $2q-m$ independent (real) equations $f_\alpha(z) = 0$, $z \in \mathbb{C}^q$. A complex vector X at $p \in \mathbb{C}^q$ is given by

$$X = \sum_{j=1}^{q} \left(a_j \frac{\partial}{\partial z^j} + b_j \frac{\partial}{\partial \overline{z}^j}\right).$$

X is said to be <u>holomorphic</u> if the b_j's are all zero and <u>antiholomorphic</u> if the a_j's are all zero. For $p \in M^m$, we say X is <u>tangent</u> to M^m at p if $Xf_\alpha = 0$, $1 \leq \alpha \leq 2q-m$. In particular the <u>holomorphic tangent space</u> HT_pM^m at $p \in M^m$ is defined by

$$HT_pM^m = \{X = \sum_{j=1}^{q} a_j \frac{\partial}{\partial z^j} \mid (Xf_\alpha)(p) = 0,\ 1 \leq \alpha \leq 2q-m\}.$$

It is easy to check that HT_pM^m is isomorphic to the maximal complex subspace contained in T_pM^m. Similarly we define the <u>antiholomorphic tangent space</u> $\overline{HT_pM^m}$ by

$$\overline{HT_pM^m} = \{X = \sum_{j=1}^{q} b_j \frac{\partial}{\partial \overline{z}^j} \mid (Xf_\alpha)(p) = 0,\ 1 \leq \alpha \leq 2q-m\}.$$

If $X_k(p) = \sum_{j=1}^{q} a_k^j(p) \frac{\partial}{\partial z^j}$ is a basis of HT_pM^m, then $\overline{X}_k(p) = \sum_{j=1}^{q} \overline{a_k^j(p)} \frac{\partial}{\partial \overline{z}^j}$ is a basis of $\overline{HT_pM^m}$.

Let M^m be a C^∞ manifold and $T_{\mathbb{C}}M^m$ the complexified tangent bundle. Suppose HM^m is a complex C^∞ subbundle of $T_{\mathbb{C}}M^m$ of complex dimension ℓ and $\mathcal{H}M^m$ the sheaf of germs of C^∞ sections of HM^m. A C-R <u>manifold of real dimension</u> m <u>and</u> C-R <u>dimension</u> ℓ is a <u>pair</u> (M^m, HM^m) such that $[\mathcal{H}_pM^m, \mathcal{H}_pM^m] \subset \mathcal{H}_pM^m$ and $\mathcal{H}_pM^m \cap \overline{\mathcal{H}_pM^m} = 0$, $p \in M^m$.

For example if M^m is a C^∞ real submanifold imbedded and locally closed in $\mathbb{C}^q$ such that dim HT_pM^m is constant on M^m, then M^m is a C-R manifold with C-R dimension = dim HT_pM^m. For if $X = \sum_{j=1}^{q} a_j\, \partial/\partial z^j \in HT_pM^m$, set $a_j = a_j' + \sqrt{-1}\, a_j''$, $z^j = x^j - \sqrt{-1}\, y^j$ and $V = \frac{1}{2} \sum_{j=1}^{q} (a_j'\, \partial/\partial x^j + a_j''\, \partial/\partial y^j)$. Then $X = V - \sqrt{-1}\, JV$ where J is the almost complex structure on $\mathbb{C}^q (J \frac{\partial}{\partial x^j} = \frac{\partial}{\partial y^j})$ and $V \in T_pM^m$. Conversely if $V \in T_pM^m$, $V - \sqrt{-1}\, JV \in HT_pM^m$. Thus if $X = V - \sqrt{-1}\, JV$ and $Y = W - \sqrt{-1}\, JW$ are in HT_pM^m,

$$
\begin{aligned}
[X,Y] &= [V,W] - [JV,JW] - \sqrt{-1}\,[JV,W] - \sqrt{-1}\,[V,JW] \\
&= -J[JV,W] - J[V,JW] - \sqrt{-1}\,[JV,W] - \sqrt{-1}\,[V,JW] \\
&= -J([JV,W] - \sqrt{-1}\,J[JV,W]) - J([V,JW] - \sqrt{-1}\,J[V,JW]) \\
&\in HT_pM^m
\end{aligned}
$$

where we have used the vanishing of $\overline{[J,J]}$ on $\mathbb{C}^q$. It is also easy to check that $HT_pM^m \cap \overline{HT_pM^m} = 0$.

For $M^m = \psi(U)$ and $r = \ell$, the vector fields P_k and $\overline{P}_k$ at $x \in U$ span vector spaces H_x and $\overline{H}_x$, each of dimension ℓ and such that $H_x \cap \overline{H}_x = 0$. Thus since the set of the P_k's is involutive M^m is a C-R manifold.

Summing up the paper of Andreotti and Hill we see that a C-R structure is locally equivalent to the data of an open set $U \subset \mathbb{R}^m$ and ℓ vector fields $P_k = \sum_{j=1}^{m} c_k^j(x) \frac{\partial}{\partial x^j}$, $1 \leq k \leq \ell$ such that the system is involutive and the 2ℓ vectors $P_k, \overline{P}_k$ are linearly independent on U. Thus if the coefficients of the P_k's are real analytic on a C-R manifold M^m of C-R dimension ℓ, we have locally $N = m - \ell$ complex characteristic coordinates.

Now let M^{2n+1} be an almost contact manifold with structure tensors φ, ξ, η. Since $\varphi\xi = 0$ and $\varphi^2 = -I + \eta \otimes \xi$, the eigenvalues of φ are 0 and $\pm\sqrt{-1}$, $\pm\sqrt{-1}$ each having multiplicity n, that is φ acts as an almost complex structure on the distribution D defined by $\eta = 0$. Thus the complexification of D in $T_{\mathbb{C}}M^{2n+1}$ is decomposable at $m \in M^{2n+1}$ into $D'_m \oplus D''_m$ where $D'_m = \{X - \sqrt{-1}\,\varphi X \mid X \in D_m\}$ and $D''_m = \{X + \sqrt{-1}\,\varphi X \mid X \in D_m\}$. Note also that for any vector of the form $X - \sqrt{-1}\,\varphi X$, $X \in M_m^{2n+1}$, $\varphi(-\varphi(X - \sqrt{-1}\,\varphi X)) = \sqrt{-1}\,(-\varphi(X - \sqrt{-1}\,\varphi X))$ so that $-\varphi(X - \sqrt{-1}\,\varphi X) \in D'_m$. We now show that if the structure (φ, ξ, η) is normal, (M^{2n+1}, D') is a C-R manifold (Ianus [29]).

<u>Theorem</u>. Let M^{2n+1} be a normal almost contact manifold, then (M^{2n+1}, D') is a C-R manifold.

<u>Proof</u>. Note that $\overline{D}'_m = D''_m$ and $D'_m \cap D''_m = 0$. Thus it suffices to show that for $X, Y \in D$, $[X - \sqrt{-1}\,\varphi X, Y - \sqrt{-1}\,\varphi Y] \in D'$. Now for X, Y in D, $[\varphi, \varphi](X, Y) + 2d\eta(X, Y)\xi = 0$ becomes

$$-[X,Y] + [\varphi X, \varphi Y] - \varphi[\varphi X, Y] - \varphi[X, \varphi Y] = 0 .$$

Also by normality $N^{(2)} = 0$, so that $(\mathfrak{L}_{\varphi X}\eta)(Y) - (\mathfrak{L}_{\varphi Y}\eta)(X) = 0$ which for X, Y in D becomes

$$\eta([\varphi X, Y] - [\varphi Y, X]) = 0.$$

Therefore

$$
\begin{aligned}
&[X - \sqrt{-1}\,\varphi X,\ Y - \sqrt{-1}\,\varphi Y] \\
&= [X,Y] - [\varphi X, \varphi Y] - \sqrt{-1}\,[\varphi X, Y] - \sqrt{-1}\,[X, \varphi Y] \\
&= -\varphi[\varphi X, Y] - \varphi[X, \varphi Y] + \sqrt{-1}\,\varphi^2[\varphi X, Y] - \sqrt{-1}\,\eta([\varphi X, Y])\xi \\
&\qquad + \sqrt{-1}\,\varphi^2[X, \varphi Y] - \sqrt{-1}\,\eta([X, \varphi Y])\xi \\
&= -\varphi([\varphi X, Y] - \sqrt{-1}\,\varphi[\varphi X, Y]) - \varphi([X, \varphi Y] - \sqrt{-1}\,\varphi[X, \varphi Y]) \\
&\in D'
\end{aligned}
$$

as desired.

Other discussions of normality as an integrability condition can be found in Kurita [35] and Sasaki and Hsu [62] (see also Sasaki [59]).

As a further remark on C-R manifolds (M,HM) let P denote the projection to the distribution $V = \mathrm{Re}(H \oplus \bar{H})$ and J the almost complex structure on V. Then setting $f = JP$ we obtain the following result of Ianus [29].

<u>Proposition</u>. Every C-R manifold carries an f-structure, i.e. a tensor field f of type (1,1) such that $f^3 + f = 0$ (Yano [82]).

We also have the following result of Yano and Ishihara [85] (see also Ludden [38]).

<u>Proposition</u>. Let J be an almost complex structure on a manifold M and N a C^∞ submanifold such that $\dim(T_pN \cap JT_pN)$ is a constant, say s on N. Then N has an f-structure of rank s.

3. K-Contact Structures

Let M^{2n+1} be a contact metric manifold, that is η is a contact form and (φ,ξ,η,g) is an associated almost contact metric structure. In this section we briefly consider the case where the characteristic vector field ξ generates a group of isometries of g; that is ξ is a Killing vector field with respect to g and we call such a contact metric structure a K-contact structure. The first basic property of a K-contact structure is that

$$\nabla_X\xi = -\varphi X \tag{8}$$

where ∇ is the Riemannian connection of g. This is easily seen as

$$\begin{aligned} g(X,\varphi Y) &= d\eta(X,Y) = \frac{1}{2}((\nabla_X\eta)(Y) - (\nabla_Y\eta)(X)) \\ &= \frac{1}{2}(g(\nabla_X\xi,Y) - g(\nabla_Y\xi,X)) \\ &= g(\nabla_X\xi,Y)\ . \end{aligned}$$

Conversely, as φ is a skew-symmetric operator, a contact metric structure satisfying equation (8) is K-contact.

In section 1 we saw that a contact metric structure is K-contact if and only if $\mathfrak{L}_\xi\varphi(= N^{(3)}) = 0$.

We now give an interesting curvature property of K-contact manifolds (Hatakeyama, Ogawa and Tanno [27]) and in turn other characterizations of K-contact manifolds.

Proposition. Let M^{2n+1} be a K-contact manifold with structure tensors (φ,ξ,η,g). Then the sectional curvature of any plane section containing ξ is equal to 1.

Proof. Let X be a unit vector field orthogonal to ξ and R the curvature tensor of g. Then

$$\begin{aligned} R_{\xi X}\xi &= \nabla_\xi\nabla_X\xi - \nabla_X\nabla_\xi\xi - \nabla_{[\xi,X]}\xi \\ &= -\nabla_\xi\varphi X + \varphi[\xi,X] \\ &= -\varphi\nabla_X\xi \\ &= \varphi^2X = -X \end{aligned}$$

and hence $g(R_{\xi X}X,\xi) = 1$.

As a corollary, we see that on a K-contact manifold of dimension $2n+1$, the Ricci curvature in the direction ξ is equal to $2n$. The converse of this is also true giving a characterization of K-contact manifolds [7].

Theorem. A contact metric manifold M^{2n+1} is a K-contact manifold if and only if the Ricci curvature in the direction of the characteristic vector field ξ is equal to $2n$.

Proof. We have just noted the necessity so it remains to prove the sufficiency. Setting $Y = \xi$ in equation (7) and using the symmetry of $\mathfrak{L}_\xi\varphi$, we obtain a general formula for $\nabla_X\xi$ for a contact metric structure.

$$\begin{aligned} 2g((\nabla_X\varphi)\xi,Z) &= g(\varphi^2[\xi,Z] - \varphi[\xi,\varphi Z],X) - 2d\eta(\varphi Z,X) \\ &= -g(\varphi(\mathfrak{L}_\xi\varphi)Z,\varphi X) - 2g(\varphi Z,\varphi X) \end{aligned}$$

$$= - g((\mathfrak{L}_\xi\varphi)Z,X) - 2g(Z,X) + 2\eta(Z)\eta(X)$$

$$= - g((\mathfrak{L}_\xi\varphi)X,Z) - 2g(X,Z) + 2g(\eta(X)\xi,Z),$$

that is

$$- \varphi\nabla_X\xi = - \frac{1}{2}(\mathfrak{L}_\xi\varphi)X - X + \eta(X)\xi .$$

Applying φ we have

$$\nabla_X\xi = - \frac{1}{2}\varphi(\mathfrak{L}_\xi\varphi)X - \varphi X . \tag{9}$$

For convenience we define an operator h by $h = \frac{1}{2}\mathfrak{L}_\xi\varphi = \frac{1}{2}N^{(3)}$ equation (9) then becomes

$$\nabla_X\xi = - \varphi X - \varphi hX . \tag{10}$$

Using equation (9) we compute $R_{\xi X}\xi$.

$$\begin{aligned} R_{\xi X}\xi &= \nabla_\xi\nabla_X\xi - \nabla_{[\xi,X]}\xi \\ &= -\frac{1}{2}\varphi\nabla_\xi([\xi,\varphi X] - \varphi[\xi,X]) - \varphi\nabla_\xi X + \frac{1}{2}\varphi(\mathfrak{L}_\xi\varphi)[\xi,X] + \varphi[\xi,X] \\ &= \frac{1}{2}\varphi\nabla_\xi\nabla_{\varphi X}\xi + \frac{1}{2}\nabla_\xi\nabla_X\xi - \frac{1}{2}\varphi\nabla_{\varphi[\xi,X]}\xi - \frac{1}{2}\nabla_{[\xi,X]}\xi - \varphi\nabla_X\xi . \end{aligned}$$

Therefore

$$\begin{aligned} \frac{1}{2} R_{\xi X}\xi &= \frac{1}{2}\varphi(\nabla_\xi\nabla_{\varphi X}\xi - \nabla_{[\xi,\varphi X]}\xi) + \frac{1}{2}\varphi\nabla_{(\mathfrak{L}_\xi\varphi)X}\xi - \varphi\nabla_X\xi \\ &= \frac{1}{2}\varphi R_{\xi\varphi X}\xi + \frac{1}{2}\varphi(-\frac{1}{2}\varphi(\mathfrak{L}_\xi\varphi)^2X - \varphi(\mathfrak{L}_\xi\varphi)X \\ &\quad - \frac{1}{2}(\mathfrak{L}_\xi\varphi)X - X + \eta(X)\xi \end{aligned}$$

and hence

$$\frac{1}{2}(R_{\xi X}\xi - \varphi R_{\xi\varphi X}\xi) = h^2X + \varphi^2X . \tag{11}$$

Now let $\{X_i, X_{n+i} = \varphi X_i, \xi\}$ $i = 1,\cdots, n$ be a φ-basis. Then as $h\xi = 0$ and $\varphi\xi = 0$, taking the inner product of equation (11) with X belonging to the φ-basis and summing, we see that

$$\operatorname{tr} h^2 = 2n - g(Q\xi, \xi)$$

where Q is the Ricci curvature operator. Thus if the Ricci curvature in the direction ξ is equal to $2n$, we have that $\operatorname{tr} h^2 = 0$. Since h is a symmetric operator, its eigenvalues are real, and hence the eigenvalues of h^2 are non-negative. Thus $\operatorname{tr} h^2 = 0$ implies that $h = \frac{1}{2} \mathfrak{L}_\xi \varphi = 0$ and so the contact metric structure on M^{2n+1} is K-contact.

We remark that on a K-contact manifold we actually have $Q\xi = 2n\xi$. Since ξ is Killing, $\mathfrak{L}_\xi$ commutes with covariant differentiation, i.e. in terms of local coordinates $\mathfrak{L}_\xi \left\{ {\alpha \atop \beta\gamma} \right\} = \nabla_\beta \nabla_\gamma \xi^\alpha + R_{\delta\beta\gamma}{}^\alpha \xi^\delta = 0$ or equivalently

$$\nabla_X \nabla_Y \xi - \nabla_{\nabla_X Y} \xi = R_{X\xi} Y .$$

Thus on a K-contact manifold

$$(\nabla_X \varphi) Y = R_{\xi X} Y .$$

Now let X be a unit vector orthogonal to ξ, then from equation (7) we have

$$2g((\nabla_X\varphi)X,Z) = g(N^{(1)}(X,Z),\varphi X) + 2\eta(Z)$$
$$= g(-[X,Z],\varphi X) + g([\varphi X,\varphi Z],\varphi X)$$
$$-g([\varphi X,Z],X] - g([X,\varphi Z],X) + 2\eta(Z).$$

Taking X belonging to a φ-basis and summing we have $Q\xi = 2n\xi$.

We can also give another type of converse to the proposition producing the contact form from purely Riemannian hypotheses. The result is also due to Hatakeyama, Ogawa and Tanno [27].

Proposition. Let M^{2n+1} be a Riemannian manifold admitting a unit Killing vector field ξ such that $R_{\xi X}\xi = -X$ for all vector fields X orthogonal to ξ (equivalently all sectional curvatures of plane sections containing ξ are equal to 1). Then M^{2n+1} is a K-contact manifold.

Proof. Define a 1-form η and a tensor field φ of type (1,1) by $\eta(X) = g(X,\xi)$ and $\varphi X = -\nabla_X\xi$. Since ξ is a unit Killing vector field, $\nabla_\xi\xi = 0$ and hence $\varphi\xi = 0$. Again since ξ is Killing,

$$\nabla_X\nabla_Y\xi - \nabla_{\nabla_X Y}\xi = R_{X\xi}Y$$

and so for any vector field X orthogonal to ξ

$$\varphi^2 X = \nabla_{\nabla_X\xi}\xi = R_{\xi X}\xi = -X$$

and therefore $\varphi^2 = -I + \eta\otimes\xi$. Moreover

$$d\eta(X,Y) = \frac{1}{2}(g(\nabla_X \xi, Y) - g(\nabla_Y \xi, X))$$
$$= - g(\nabla_Y \xi, X)$$
$$= g(X, \varphi Y)$$

and $g(\varphi X, \varphi Y) = - g(X, \varphi^2 Y) = g(X,Y) - \eta(X)\eta(Y)$ and hence (φ, ξ, η, g) is a K-contact structure.

Finally we prove a theorem of Tachibana on compact K-contact manifolds [67].

Theorem. On a compact K-contact manifold M^{2n+1}, any harmonic 1-form ω is orthogonal to η, that is $\iota(\eta)\omega = 0$.

Proof. Recall that for a Killing vector field ξ with covariant form η, $\Delta\eta = - 2\eta \circ Q$ and that on a compact orientable Riemannian manifold a harmonic 1-form ω watisfies $\mathfrak{L}_\xi \omega = 0$ (see e.g. Goldberg [19]). From $\mathfrak{L}_\xi \omega = 0$ we have $d(\iota(\xi)\omega) = 0$, that is $\omega(\xi)$ is constant on M^{2n+1}. Define a 1-form β by $\omega = \omega(\xi)\eta + \beta$, then $\beta(\xi) = 0$ and

$$\Delta\beta = - \omega(\xi)\Delta\eta = 2\omega(\xi)\eta \circ Q = 4n\omega(\xi)\eta$$

since $Q\xi = 2n\xi$ as we have seen. Thus the global scalar product $(\Delta\beta, \beta) = 0$ $((\alpha,\beta) = \int_{M^{2n+1}} \alpha \wedge * \beta)$, so that $0 = ((d\delta + \delta d)\beta, \beta) = (\delta\beta, \delta\beta) + (d\beta, d\beta)$. Therefore β is harmonic and hence $\omega(\xi) = 0$.

4. Regular Contact Manifolds.

In example I.2.H we gave the Boothby-Wang fibration of a compact regular contact manifold M^{2n+1} as a principal circle bundle over a symplectic manifold M^{2n}. Since M^{2n} carries a global symplectic form Ω, there exist a Riemannian metric G and a tensor field J of type (1,1) such that (J,G) is an almost Kähler structure on M^{2n} with Ω as its fundamental 2-form. Let η denote the contact form on M^{2n+1} with $d\eta = \pi^*\Omega$ and ξ its characteristic vector field and define a tensor field φ by $\varphi X = \tilde{\pi} J\pi_* X$, $X \in T_m M^{2n+1}$. Then $\varphi^2 = -I + \eta \otimes \xi$ and hence (φ, ξ, η) is an almost contact structure. Define a Riemannian metric g on M^{2n+1} by $g = \pi^* G + \eta \otimes \eta$. Clearly ξ is a unit Killing vector field with respect to g. Moreover $g(X,\varphi Y) = G(\pi_* X, J\pi_* Y) \circ \pi = \Omega(\pi_* X, \pi_* Y) \circ \pi = \pi^*\Omega(X,Y) = d\eta(X,Y)$ and similarly $g(\varphi X, \varphi Y) = g(X,Y) - \eta(X)\eta(Y)$. Thus we see that a compact regular contact manifold carries a K-contact structure (Hatakeyama [26]).

We can also give a topological result on compact regular contact manifolds. Since the characteristic class of the principal circle bundle $\pi: M^{2n+1} \to M^{2n}$ is $[\Omega] \in H^2(M^{2n}, \mathbb{Z})$ the bundle is non-trivial and hence the Gysin sequence becomes

$$0 \longrightarrow H^1(M^{2n}, \mathbb{R}) \xrightarrow{\pi^*} H^1(M^{2n+1}, \mathbb{R}) \longrightarrow H^0(M^{2n}, \mathbb{R})$$

$$\xrightarrow{L} H^2(M^{2n}, \mathbb{R}) \longrightarrow \cdots$$

where L is left exterior multiplication by Ω. Thus L is an isomorphism of $H^0(M^{2n},\mathbb{R})$ into $H^2(M^{2n},\mathbb{R})$ and therefore the map π^* is an onto isomorphism giving the following theorem of Tanno [71].

<u>Theorem</u>. Let $\pi: M^{2n+1} \to M^{2n}$ denote the Boothby-Wang fibration of a compact regular contact manifold M^{2n+1}. Then the first Betti numbers of M^{2n+1} and M^{2n} are equal.

In [71], Tanno also proves this result by use of harmonic forms on M^{2n+1} considered as a K-contact manifold. As seen in the theorem of Tachibana in the last section harmonic 1-forms are orthogonal to η and projectable and one can check that the projected form is harmonic. Conversely the pullback of a harmonic form is shown to be harmonic.

It is an open question whether or not the five dimensional torus T^5 can carry a contact structure. If T^5 does carry a contact structure it cannot be regular.

<u>Theorem</u>. No torus T^{2n+1} can carry a regular contact structure.

<u>Proof</u>. If T^{2n+1} admitted a regular contact structure, T^{2n+1} would be a principal circle bundle over a symplectic manifold M^{2n} by the Boothby-Wang fibration. We have just seen that the first Betti number of the base $b_1(M^{2n})$ is equal to $b_1(T^{2n+1}) = 2n + 1$. On the other hand we have the homotopy sequence of the bundle

$$0 \to \pi_2(M^{2n}) \to \pi_1(S^1) \to \pi_1(T^{2n+1}) \to \pi_1(M^{2n}) \to 0$$

since $\pi_2(T^{2n+1}) = 0$. Now consider the universal covering space $\mathbb{R}^{2n+1}$ of T^{2n+1} and the lift of the fibration to $\mathbb{R}^{2n+1}$; each circle lifts to a line and hence the fibration of T^{2n+1} by circles has no null-homotopic fibres. Thus the map from $\pi_1(S^1)$ into $\pi_1(T^{2n+1})$ is non-trivial and hence $\pi_2(M^{2n}) = 0$. Then by the exactness of the sequence $\pi_1(M^{2n}) = \mathbb{Z} \oplus \cdots \oplus \mathbb{Z}/\mathbb{Z}$ and hence $b_1(M^{2n}) = 2n$, a contradiction.

The theorem is not, however, suggestive of a negative conjecture concerning the problem of a contact structure on T^5, since the theorem also says that T^3 cannot carry a regular contact structure though it carries a non-regular one (example I.2.C). In section VI.1 we will show that T^5 cannot admit a contact metric structure with a flat associated metric, but T^3 does carry a flat associated metric to its contact structure.

5. Sasakian Manifolds.

In section II.3 we showed that a contact manifold with contact form η carries an almost contact metric structure (φ,ξ,η,g) with $\Phi = d\eta$. This structure was referred to as an associated structure or simply as a contact metric structure. If a contact metric structure (φ,ξ,η,g) is normal, we call it a normal contact metric or a Sasakian structure.

A Sasakian structure is in some sense an analogue of a Kähler structure on an almost Hermitian manifold, i.e. the almost complex structure J is parallel with respect to the Hermitian metric. This point of view is suggested in the following formulation of the Sasakian condition [61].

Theorem. An almost contact metric structure (φ,ξ,η,g) is Sasakian if and only if

$$(\nabla_X\varphi)Y = g(X,Y)\xi - \eta(Y)X \tag{12}$$

where ∇ denotes the Riemannian connection of g.

Proof. The necessity follows easily from equation (6) for if (φ,ξ,η,g) is a normal contact metric structure, $\Phi = d\eta$, $N^{(1)} = 0$ and $N^{(2)} = 0$, and hence

$$\begin{aligned} g((\nabla_X\varphi)Y,Z) &= \Phi(\varphi Y,X)\eta(Z) - \Phi(\varphi Z,X)\eta(Y) \\ &= g(X,Y)\eta(Z) - g(X,Z)\eta(Y) \\ &= g(g(X,Y)\xi - \eta(Y)X,Z) . \end{aligned}$$

Conversely if an almost contact metric structure (φ, ξ, η, g) satisfies equation (12), setting $Y = \xi$ we have $-\varphi \nabla_Y \xi = \eta(X)\xi - X$ and consequently applying φ,

$$\nabla_X \xi = -\varphi X .$$

By the skew-symmetry of φ we see that ξ is Killing and hence

$$\begin{aligned} d\eta(X,Y) &= \frac{1}{2}((\nabla_X \eta)(Y) - (\nabla_Y \eta)(X)) \\ &= g(\nabla_X \xi, Y) = -g(\varphi X, Y) = \Phi(X,Y). \end{aligned}$$

Thus (φ, ξ, η, g) is a contact metric structure and from the formula

$$[\varphi, \varphi](X,Y) = (\varphi \nabla_Y \varphi - \nabla_{\varphi Y} \varphi)X - (\varphi \nabla_X \varphi - \nabla_{\varphi X} \varphi)Y$$

for the Nijenhuis torsion in terms of a symmetric connection, we have by direct substitution from equation (12) that $[\varphi, \varphi] + 2d\eta \otimes \xi = 0$.

We noted in section 3 that for a K-contact structure $R_{\xi X}\xi = -X$. Here we give a similar lemma.

Lemma. On a Sasakian manifold for a unit vector field X orthogonal to ξ we have

$$R_{X\xi}X = -\xi .$$

Proof.
$$\begin{aligned} g(R_{X\xi}X, Y) &= -g(R_{XY}\xi, X) \\ &= -g(\nabla_X \nabla_Y \xi - \nabla_Y \nabla_X \xi - \nabla_{[X,Y]}\xi, X) \end{aligned}$$

$$= - g(-\nabla_X\varphi Y + \nabla_Y\varphi X + \varphi[X,Y],X)$$

$$= g((\nabla_X\varphi)Y - (\nabla_Y\varphi)X,X)$$

$$= - g(\eta(Y)X - \eta(X)Y,X)$$

$$= - \eta(Y) = g(-\xi,Y) .$$

Actually we have proved slightly more, namely that on a Sasakian manifold we have

$$R_{XY}\xi = \eta(Y)X - \eta(X)Y .$$

Again we have a converse similar to the one given in section 3 [27].

Theorem. Let M^{2n+1} be a Riemannian manifold admitting a unit Killing vector field ξ such that $R_{XY}\xi = g(\xi,Y)X - g(X,\xi)Y$, then M^{2n+1} is a Sasakian manifold. In particular the usual contact metric structure on an odd dimensional sphere is a Sasakian structure.

Proof. As in the corresponding result in section 3 set $\eta(X) = g(X,\xi)$ and $\varphi X = -\nabla_X\xi$. Then we know that (φ,ξ,η,g) is a K-contact structure. Moreover since ξ is Killing

$$\nabla_X\nabla_Y\xi - \nabla_{\nabla_X Y}\xi = R_{X\xi}Y$$

from which

$$(\nabla_X\varphi)Y = R_{\xi X}Y$$

and therefore

$$g((\nabla_X\varphi)Y,Z) = g(R_{\xi X}Y,Z) = g(R_{YZ}\xi,X)$$

$$= g(\eta(Z)Y - \eta(Y)Z,X)$$

which is just

$$(\nabla_X\varphi)Y = g(X,Y)\xi - \eta(Y)X$$

completing the proof.

Sasakian manifolds have many properties analogous to Kähler manifolds, for example in the lemma of section V.1 we shall see that for vector fields X,Y,Z and W orthogonal to ξ on a Sasakian manifold

$$g(R_{\varphi X\varphi Y}\varphi Z,\varphi W) = g(R_{XY}Z,W) .$$

Thus choosing a φ-basis $\{X_i, X_{n+i} = \varphi X_i, \xi\}$ we have for the Ricci tensor S and X and Y orthogonal to ξ

$$S(\varphi X,\varphi Y) = \sum_{A=1}^{2n} g(R_{\varphi X X_A}X_A,\varphi Y) + g(R_{\varphi X\xi}\xi,\varphi Y)$$

$$= \sum_{A=1}^{2n} g(R_{\varphi X\varphi X_A}\varphi X_A,\varphi Y) + g(X,Y)$$

$$= S(X,Y) .$$

We already know that $Q\xi = 2n\xi$ and hence the Ricci operator Q commutes with φ on a Sasakian manifold.

In the Sasakian case we can continue the theorem of Tachibana in the last section [67].

Theorem. The first Betti number of a compact Sasakian manifold M^{2n+1} is zero or even.

Proof. Recall that the Laplacian of a 1-form u is given in terms of local coordinates $(x^1,\cdots, x^{2n+1})$ by

$$(\Delta u)_\alpha = g^{\nu\mu}\nabla_\nu\nabla_\mu u_\alpha - R_\alpha{}^\mu u_\mu$$

where the summation convention is employed on repeated indices and $R_\lambda{}^\alpha$ denote the components of the Ricci operator Q. Suppose that u is a harmonic 1-form on M^{2n+1} and define a 1-form $\tilde{u}$ by $\tilde{u} = u\circ\varphi$. Then

$$\begin{aligned} g^{\mu\nu}\nabla_\nu\nabla_\mu\tilde{u}_\lambda &= g^{\mu\nu}\nabla_\nu[(g_{\mu\lambda}\xi^\alpha - \eta_\lambda\delta_\mu^\alpha)u_\alpha + \varphi_\lambda{}^\alpha\nabla_\mu u_\alpha] \\ &= g^{\mu\nu}[g_{\mu\lambda}\nabla_\nu(\xi^\alpha u_\alpha) - (\nabla_\nu\eta_\lambda)u_\mu - \eta_\lambda\nabla_\nu u_\mu \\ &\quad + (g_{\nu\lambda}\xi^\alpha - \eta_\lambda\delta_\nu^\alpha)\nabla_\mu u_\alpha + \varphi_\lambda{}^\alpha\nabla_\nu\nabla_\mu u_\alpha] \\ &= 2\nabla_\nu(\xi^\alpha u_\alpha) - 2\eta_\lambda\nabla_\nu u^\nu + \varphi_\lambda{}^\alpha g^{\mu\nu}\nabla_\nu\nabla_\mu u_\alpha \\ &= \varphi_\lambda{}^\alpha R_\alpha{}^\mu u_\mu = R_\lambda{}^\alpha\varphi_\alpha{}^\mu u_\mu = R_\lambda{}^\alpha\tilde{u}_\alpha \end{aligned}$$

where we have used the Theorem of Tachibana in section 3 $(\xi^\alpha u_\alpha = 0)$, $\nabla_\nu\xi^\alpha = -\varphi_\nu{}^\alpha$, $\delta u = 0$ and $Q\varphi = \varphi Q$. Thus we see that $\tilde{u}$ is harmonic, but u and $\tilde{u}$ are independent so the first Betti number of M^{2n+1} is zero or even.

In the 1960's a great deal of work was done on the topology of compact Sasakian manifolds. In addition to the papers already mentioned (Tachibana [67], Tanno [71])we have for example that the p-th Betti number b_p is even for p odd and $1 \leq p \leq n$ and by duality b_p is even for p even and $n+1 \leq p \leq 2n$, (Blair and Goldberg [8], Fujitani [18]). The fundamental group has been studied by Blair and Goldberg [8] and Harada [24, 25]. Considerable attention has been given to vanishing of the second Betti number under some curvature restrictions as well as being isometric to a sphere under stronger conditions. A compact Sasakian manifold of strictly positive curvature has vanishing second Betti number (Moskal [43], Tanno [72], Goldberg [20] in the regular case and Goldberg in the regular case with non-negative sectional curvature [22]). A compact, simply connected Sasakian-Einstein space of strictly positive curvature is isometric to a unit sphere (Moskal [43]). A compact, simply connected Sasakian symmetric space is isometric to a sphere (Goldberg [22], Okumura [47]). A conformally flat Sasakian manifold of dimension ≥ 5 is of constant curvature (Okumura [47]). Pinching theorems have been obtained by Tanno including an analogue of holomorphic pinching [72].

Much of the development of Chapter II and the present chapter has been analogous to the theory of almost complex manifolds. Before turning to our examples let us summarize the various structures involved in the two theories.

Let M^{2n} be an almost complex manifold with structure tensor J, G an Hermitian metric, Ω the fundamental

2-form and ∇ the Riemannian connection of G. Then note the following schematic array of structures.

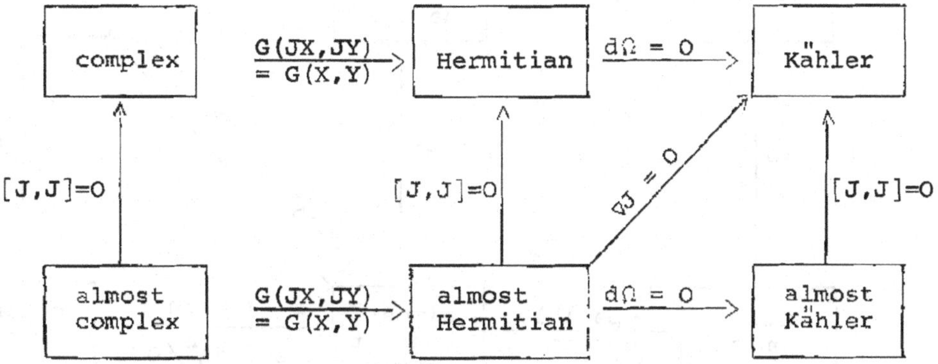

Recall that S^6 carries an almost Hermitian structure that is neither Hermitian nor almost Kähler, the well known Calabi-Eckmann manifolds $S^{2p+1} \times S^{2q+1}$, $p,q \geq 1$ are Hermitian manifolds which are not Kählerian, the tangent bundle to a non-flat Riemannian manifold carries an almost Kähler structure which is not Kählerian (see e.g. section VII.1) and there are many well known Kähler manifolds, e.g. $\mathbb{C}^n$, $P\mathbb{C}^n$, etc.

The corresponding diagram for almost contact manifolds is found on the following page, the notion of a K-contact manifold being intermediate between a contact metric manifold and a Sasakian manifold. We have already seen that S^5 carries an almost contact metric structure which is not a contact metric structure and in example B we shall see that it is not normal; similarly $S^6 \times \mathbb{R}$ as a special case of example II.4.D is a non-normal almost contact manifold.

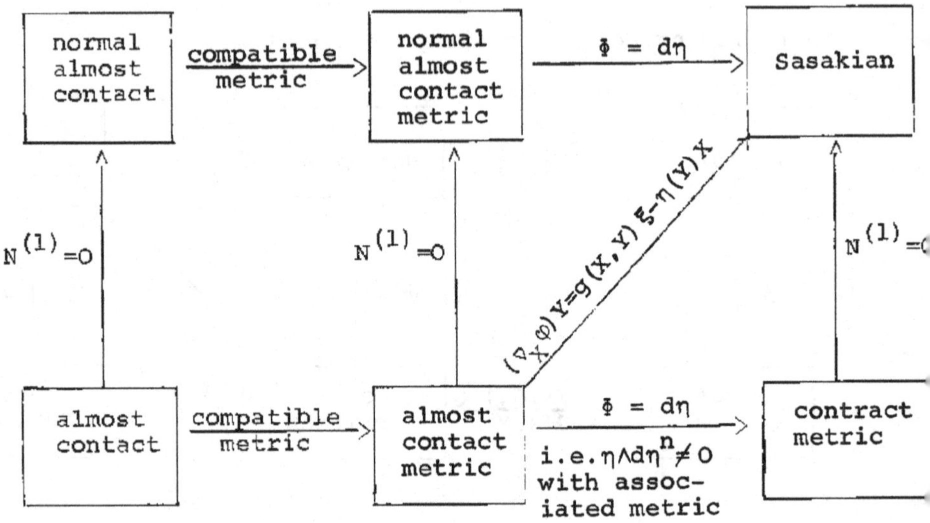

On the other hand choosing M^{2n} to be a complex manifold in example II.4.D we have a normal almost contact structure with $d\eta = 0$. By the topological result on the first Betti number of a Sasakian manifold we see that the three dimensional torus cannot carry a Sasakian structure. In section VII.2 we shall also see that the tangent sphere bundles are not in general Sasakian. Finally the odd-dimensional spheres are Sasakian manifolds as well as Examples A,C,D and E below.

6. Examples.

A. $\mathbb{R}^{2n+1}$. In Example II.4.A. we gave explicitly an associated almost contact metric structure (φ,ξ,η,g) to the usual contact structure $\eta = \frac{1}{2}(dz - \sum_{i=1}^{n} y^i dx^i)$ on $\mathbb{R}^{2n+1}$, (x^i, y^i, z) being cartesian coordinates. From the matrix expression for φ it is easy to check that $[\varphi,\varphi] + 2d\eta \otimes \xi = 0$ and hence that this contact metric structure is normal.

In view of our discussion in section 2 we make one further remark about this example. Let $\eta = \frac{1}{2}(dz - y\,dx)$ be the usual contact structure on $\mathbb{R}^3$, then the contact distribution D is spanned by $X_1 = \frac{\partial}{\partial x} + y\frac{\partial}{\partial z}$ and $X_2 = \frac{\partial}{\partial y}$. Thus any vector $V \in D$ is given by $V = aX_1 + bX_2 = \begin{pmatrix} a \\ b \\ ay \end{pmatrix}$. The matrix of φ is

$$\begin{pmatrix} 0 & 1 & 0 \\ -1 & 0 & 0 \\ 0 & y & 0 \end{pmatrix}$$

and therefore the distribution D' in the complexification of D is spanned by $P = V - \sqrt{-1}\,\varphi V = (a - \sqrt{-1}\,b)\begin{pmatrix} 1 \\ \sqrt{-1} \\ y \end{pmatrix}$ and D'' is spanned by $\bar{P} = V + \sqrt{-1}\,\varphi V$. The condition for a complex characteristic coordinate ζ is

$$P\zeta = (a - \sqrt{-1}\,b)\frac{\partial\zeta}{\partial x} + (a - \sqrt{-1}\,b)\sqrt{-1}\,\frac{\partial\zeta}{\partial y} + (a - \sqrt{-1}\,b)\,y\,\frac{\partial\zeta}{\partial z} = 0$$

which has as a general solution

$$\zeta = f(y - \sqrt{-1}\, x,\ z + \sqrt{-1}\, \frac{y^2}{2})$$

where f is any smooth function of two complex variables.

B. A non-normal almost contact structure on S^5.

In example II.4.C we saw that S^5 inherits from the almost Hermitian structure on S^6 an almost contact metric structure different than the standard one. Recall that the almost Hermitian structure (J,G) on S^6 induced by the vector product on $\mathbf{R}^7$ is nearly Kählerian, that is $(\tilde{\nabla}_X J)X = 0$ for all vector fields X where $\tilde{\nabla}$ is the Riemannian connection of G. The geometric meaning of the nearly Kähler condition is that geodesics are holomorphically planar curves (a curve γ on an almost Hermitian manifold is holomorphically planar if the holomorphic section determined by its tangent field is parallel along the curve; thus for geodesics on a nearly Kähler manifold $\tilde{\nabla}_{\gamma_*} J\gamma_* = (\tilde{\nabla}_{\gamma_*} J)\gamma_* = 0)$.

We will show that the induced almost contact metric structure (φ, ξ, η, g) on S^5 satisfies a similar condition, namely $(\nabla_X \varphi)X = 0$ (equivalently $(\nabla_X \varphi)Y + (\nabla_Y \varphi)X = 0)$. This is an immediate consequence of the following theorem [5].

Theorem. Let $\iota\colon M^{2n+1} \to M^{2n+2}$ be a C^∞ orientable hypersurface of a nearly Kähler manifold. Then the induced almost contact metric structure (φ,ξ,η,g) satisfies $(\nabla_X\varphi)X = 0$ if and only if the second fundamental form h is proportional to $\eta\otimes\eta$.

Proof. Using the Gauss equation $\tilde{\nabla}_{\iota_*X}\iota_*Y = \iota_*\nabla_XY + h(X,Y)C$, we have

$$(\nabla_X\Phi)(Y,Z) = (\tilde{\nabla}_{\iota_*X}\Omega)(\iota_*Y,\iota_*Z) + h(X,Y)\eta(Z) - h(X,Z)\eta(Y) \tag{13}$$

where Ω is the fundamental two form of the nearly Kähler structure. Interchanging X and Z and adding we obtain

$$(\nabla_X\Phi)(Y,Z) + (\nabla_Z\Phi)(Y,X) = -2\eta(Y)h(X,Z) + \eta(Z)h(X,Y) + \eta(X)h(Z,Y).$$

Clearly if h is proportional to $\eta\otimes\eta$, $(\nabla_X\varphi)X = 0$. Conversely if $(\nabla_X\varphi)X = 0$,

$$0 = -2\eta(Y)h(X,Z) + \eta(Z)h(X,Y) + \eta(X)h(Z,Y).$$

Setting $Y=\xi$ gives $2h(X,Z) = \eta(Z)h(X,\xi) + \eta(X)h(Z,\xi)$, but now setting $X=\xi$ we have $h(\xi,Z) = h(\xi,\xi)\eta(Z)$ and consequently

$$h(X,Z) = h(\xi,\xi)\eta(X)\eta(Z).$$

An almost contact metric structure (φ,ξ,η,g) with $(\nabla_X\varphi)X = 0$ is called a <u>nearly cosymplectic structure</u>.

There are two notions of cosymplectic structure in the literature. P. Libermann [36] defines a cosymplectic manifold to be one admitting a closed 1-form η and a closed 2-form Φ such that $\eta \wedge \Phi^n$ is a volume element. In [4] the author defined a cosymplectic structure to be a normal almost contact metric structure (φ, ξ, η, g) with Φ and η closed. Using the second definition S. I. Goldberg [21] gave a cosymplectic version of our theorem (see also Okumura [51]).

Proposition. On a nearly cosymplectic manifold ξ is a Killing vector field.

Proof. Clearly $(\nabla_\xi \varphi)\xi = 0$ and so $\varphi\nabla_\xi \xi = 0$ from which, applying φ, we have $\nabla_\xi \xi = 0$. Differentiating the compatibility condition of the metric, $g(\varphi X, \varphi Y) = g(X,Y) - \eta(X)\eta(Y)$, with respect to ξ we find

$$g((\nabla_\xi \varphi)X, \varphi Y) + g(\varphi X, (\nabla_\xi \varphi)Y) = 0 .$$

The nearly cosymplectic condition then gives

$$g((\nabla_X \varphi)\xi, \varphi Y) + g(\varphi X, (\nabla_Y \varphi)\xi) = 0$$

which easily simplifies to

$$g(\nabla_X \xi, Y) + g(\nabla_Y \xi, X) = 0 .$$

Proposition. On a normal nearly cosymplectic manifold $d\eta = 0$.

Proof. Since the structure is normal, $N^{(1)}$ and $N^{(2)}$ vanish, so setting $Y = X$ and $Z = \xi$ in equation (6) we have $d\eta(X, \varphi X) = 0$ for all vector fields X. Linearizing this we obtain

$$d\eta(X, \varphi Y) + d\eta(Y, \varphi X) = 0,$$

but by the vanishing of $N^{(2)}$ (equation (4)) $d\eta(X, \varphi Y) = - d\eta(\varphi X, Y)$ and hence $d\eta(X, \varphi Y) = 0$. Now $d\eta(X, \xi) = \frac{1}{2}(g(\nabla_X \xi, \xi) - g(\nabla_\xi \xi, X)) = 0$ so putting these together we see that $d\eta = 0$.

Turning now to S^5 as a totally geodesic hypersurface of S^6, its induced structure is nearly cosymplectic by our Theorem. We now show that this structure is not normal [5]. For suppose it were, then by our two propositions η is respectively co-closed and closed and hence harmonic, contradicting the vanishing of the first Betti number of S^5

Finally as a further justification of the name nearly cosymplectic we prove the following theorem [5].

Theorem. A normal nearly cosymplectic manifold is cosymplectic.

Proof. We have already seen that $d\eta = 0$ in this case, so returning again to equation (6) with $Y = X$ we see that $d\Phi(X, \varphi X, \varphi Z) = 0$ from which

$$d\Phi(X, \varphi Y, \varphi Z) + d\Phi(Y, \varphi X, \varphi Z) = 0.$$

Therefore $d\Phi(\xi,X,Y) = 0$ and

$$d\Phi(X,\varphi Y,\varphi Z) = d\Phi(\varphi X,Y,\varphi Z) = -\, d\Phi(\varphi X,\varphi Z,Y)$$

$$= d\Phi(X,Z,Y) = -\, d\Phi(X,Y,Z)\ .$$

Consequently equation (6) becomes

$$g((\nabla_X\varphi)Y,Z) = -\,3d\Phi(X,Y,Z)$$

$$= g((\nabla_X\varphi)Y,Z) + g((\nabla_Y\varphi)Z,X) + g((\nabla_Z\varphi)X,Y)$$

$$= 3g((\nabla_X\varphi)Y,Z)$$

and hence $\nabla_X\varphi = 0$ and then of course $d\Phi = 0$.

C. <u>Principal circle bundles</u>. In section 4 we saw that a compact regular contact manifold M^{2n+1} carries a K-contact structure (φ,ξ,η,g), the base manifold M^{2n} of the fibration of M^{2n+1} being almost Kählerian. Since $N^{(3)} = \mathfrak{L}_\xi\varphi =$

$$[\varphi,\varphi](\xi,X) + 2d\eta(\xi,X)\xi = \varphi^2[\xi,X] - \varphi[\xi,\varphi X] = 0\ .$$

Now $\varphi X = \tilde{\pi}J\pi_*X$, so for projectable horizontal vector fields X and Y

$$[\varphi,\varphi](X,Y) + 2d\eta(X,Y)\xi$$

$$= \tilde{\pi}J^2\pi_*[X,Y] + [\tilde{\pi}J\pi_*X,\tilde{\pi}J\pi_*Y] - \tilde{\pi}J\pi_*[\tilde{\pi}J\pi_*X,Y]$$

$$- \tilde{\pi}J\pi_*[X,\tilde{\pi}J\pi_*Y] + 2d\eta(X,Y)\xi$$

$$= \tilde{\pi}J^2[\pi_*X,\pi_*Y] + \tilde{\pi}[J\pi_*X,J\pi_*Y] + \eta([\tilde{\pi}J\pi_*X,\tilde{\pi}J\pi_*Y])\xi$$

$$- \tilde{\pi}J[J\pi_*X,\pi_*Y] - \tilde{\pi}J[\pi_*X,J\pi_*Y] + 2d\eta(X,Y)\xi$$

$$= \tilde{\pi}[J,J](\pi_*X,\pi_*Y) - 2(\Omega(J\pi_*X,J\pi_*Y)\circ\pi - \Omega(\pi_*X,\pi_*Y)\circ\pi)\xi$$

$$= \tilde{\pi}[J,J](\pi_*X,\pi_*Y) .$$

Thus we see that the K-contact structure (φ,ξ,η,g) is Sasakian if and only if the base manifold M^{2n} is Kählerian (Hatakeyama [26]).

Similar to the Boothby-Wang fibration of compact regular contact manifolds, A. Morimoto [42] obtained a fibration of compact normal almost contact manifolds with ξ regular. First however let $\pi: M^{2n+1} \to M^{2n}$ be a principal circle bundle over a complex manifold M^{2n} and suppose there exists a connection form η such that $d\eta = \pi^*\Psi$ where Ψ is a form of bidegree (1,1) on M^{2n}. Then we again define φ by $\varphi X = \tilde{\pi}J\pi_*X$ where J is the almost complex structure on M^{2n} and $\tilde{\pi}$ the horizontal lift with respect to η and let ξ be a vertical vector field with $\eta(\xi) = 1$. Clearly (φ,ξ,η) is an almost contact structure. Noting that $\mathfrak{L}_\xi\eta = 0$ since η is a connection form and $\mathfrak{L}_\xi\varphi = 0$ by the definition of φ, a computation of $N^{(1)}$ similar to the one just given shows that M^{2n+1} is a normal almost contact manifold [41].

Conversely we prove the following theorem of Morimoto [42] and sketch its proof since the major ideas have already been given.

<u>Theorem</u>. Let M^{2n+1} be a compact normal almost contact manifold with structure tensor (φ, ξ, η) and suppose that ξ is a regular vector field. Then M^{2n+1} is the bundle space of a principal circle bundle $\pi: M^{2n+1} \to M^{2n}$ over a complex manifold M^{2n}. Moreover η is a connection form and the 2-form Ψ on M^{2n} such that $d\eta = \pi^*\Psi$ is of bidegree (1,1).

<u>Proof</u>. In the proof of the Boothby-Wang Theorem (example I.2.H) we defined the period function λ of the vector field ξ and showed that λ was constant on M^{2n+1} which we then took to be 1. The argument (cf. Tanno [70]) required only that $\eta(\xi) = 1$ and $\mathfrak{L}_\xi\eta = N^{(4)} = 0$. Thus we again have a circle bundle structure as in the Boothby-Wang fibration with η a connection form. Now since $N^{(3)} = \mathfrak{L}_\xi\varphi = 0$, φ is projectable and we can define an almost complex structure J on M^{2n} by $JX = \pi_*\varphi\tilde{\pi}X$ where $\tilde{\pi}$ denotes the horizontal lift with respect to η. That $J^2 = -I$ is immediate and

$$\begin{aligned}
\tilde{\pi}[J,J](X,Y) &= -[\tilde{\pi}X,\tilde{\pi}Y] + \eta([\tilde{\pi}X,\tilde{\pi}Y])\xi + [\varphi\tilde{\pi}X,\varphi\tilde{\pi}Y] \\
&\quad - \eta([\varphi\tilde{\pi}X,\varphi\tilde{\pi}Y])\xi - \varphi[\varphi\tilde{\pi}X,\tilde{\pi}Y] - \varphi[\tilde{\pi}X,\varphi\tilde{\pi}Y] \\
&= [\varphi,\varphi](\tilde{\pi}X,\tilde{\pi}Y) + 2d\eta(\varphi\tilde{\pi}X,\varphi\tilde{\pi}Y)\xi \\
&= [\varphi,\varphi](\tilde{\pi}X,\tilde{\pi}Y) + 2d\eta(\tilde{\pi}X,\tilde{\pi}Y)\xi \\
&= 0,
\end{aligned}$$

the last equality following from $N^{(1)} = 0$ and the next to last from $N^{(2)} = 0$ (cf. equation (5)). Finally $\Psi(JX,JY) \circ \pi = d\eta(\varphi\widetilde{\pi X}, \varphi\widetilde{\pi Y}) = d\eta(\widetilde{\pi X}, \widetilde{\pi Y}) = \Psi(X,Y) \circ \pi$ showing that Ψ is of bidegree (1,1).

D. $M^{2n+1} \subset M^{2n+2}$. That the usual contact metric structure on the odd-dimensional sphere is Sasakian can be seen in many ways, for example the second theorem of section 5 or considering the Hopf fibration $\pi: S^{2n+1} \to P\mathbb{C}^n$ as a special case of the Boothby-Wang fibration in example C. Here we prove a theorem due to Tashiro [75] similar to the one in the cosymplectic case in example B.

Theorem. Let $\iota: M^{2n+1} \to M^{2n+2}$ be a C^∞ orientable hypersurface of a Kähler manifold. Then the induced almost contact metric structure (φ, ξ, η, g) is Sasakian if and only if the second fundamental form $h = -g + \beta\eta \otimes \eta$, β a function.

Proof. Since the ambient space is Kählerian equation (13) becomes

$$(\nabla_X \Phi)(Y,Z) = h(X,Y)\eta(Z) - h(X,Z)\eta(Y) .$$

If $h = -g + \beta\eta \otimes \eta$, we have

$$(\nabla_X \Phi)(Y,Z) = -g(X,Y)\eta(Z) + g(X,Z)\eta(Y)$$

and hence (φ, ξ, η, g) is a Sasakian structure. Conversely if M^{2n+1} is Sasakian we have

$$h(X,Y)\eta(Z) - h(X,Z)\eta(Y) = -g(X,Y)\eta(Z) + g(X,Z)\eta(Y). \quad (14)$$

Setting $X = Y = \xi$ we have $h(\xi,\xi)\eta(Z) - h(\xi,Z) = 0$; using this result and setting $Z = \xi$ in equation (14) we have

$$h(X,Y) - h(\xi,\xi)\eta(X)\eta(Y) = -g(X,Y) + \eta(X)\eta(Y)$$

or $h = -g + \beta\eta\otimes\eta$ where $\beta = h(\xi,\xi) + 1$.

A contact metric structure (φ,ξ,η,g) is said to be <u>nearly Sasakian</u> if $(\nabla_X\varphi)Y + (\nabla_Y\varphi)X = 2g(X,Y)\xi - \eta(X)Y - \eta(Y,X)$. The above theorem also holds for this structure on a hypersurface of a nearly Kähler manifold [12]. Similar to the structure on S^5 obtained in example B; consider S^5 as an umbilical hypersurface of the unit sphere S^6 at a "latitude" of 45° so that $h = -g$. Then the induced almost contact metric structure is nearly Sasakian but not Sasakian.

Other results on hypersurfaces of Kähler manifolds may be found in Okumura [49], [50] and [52].

E. <u>A non-regular Sasakian structure</u>. We have seen that a regular contact manifold admits a K-contact structure, but a K-contact structure need not be regular. In fact we can give an example of a non-regular Sasakian manifold (Tanno [74]). Consider S^3 as the set of unit quaternions $\lambda = a + bi + cj + dk \in H$, $|\lambda| = 1$. Regarding λ as a vector (a,b,c,d), the unit outer normal to S^3 is just the position vector, so the usual contact structure on S^3 is obtained by applying the usual almost complex structure, $J(a,b,c,d) = (-b,a,-d,c) = \xi$. Define $\mu: S^3 \to S^3$ by $\mu(a,b,c,d) = (a,b,d,c)$ and J^* by $J^*\lambda = \lambda i$. Then $J^*\mu = \mu J$. Let $\rho = (\alpha,\beta,\gamma,\delta) \in \mathbf{H}$ such that $\rho^s = 1$ for some integer s and consider the transformation $L_\rho: S^3 \to S^3$ where L_ρ denotes left multiplication by ρ. Now $\xi^* = J^*\lambda$ also defines a vector field ξ^* on S^3 and it is easy to check that $L_{\rho_*}\xi^* = \xi^*$ and $\mu_*\xi = \xi^*$. Let $\hat{\rho} = \mu^{-1} \circ L_\rho \circ \mu$, then $\hat{\rho}$ is an isometry on S^3 which preserves ξ.

Now let $\pi: S^3 \to P\mathbb{C}^1$ be the Hopf fibration which is just the Boothby-Wang fibration of S^3 with its usual contact structure. Since $\hat{\rho}$ is an isometry and preserves ξ, $\hat{\rho}$ induces and an automorphism $\underline{\rho}$ of the Kähler structure on $P\mathbb{C}^1$. Moreover we can easily choose ρ so that $\underline{\rho}$ is non-trivial. Now every automorphism of $P\mathbb{C}^1$ has a fixed point, so the orbits of ξ over such fixed points are invariant under $\hat{\rho}$. Thus letting Γ denote a finite group generated by $\hat{\rho}$, $M^3 = S^3/\Gamma$ is a Sasakian manifold which is not regular.

Non-regular contact manifolds have just recently become the subject of several investigations, for example Abe [1], Abe and Erbacher [2], Thomas [78].

CHAPTER V

SASAKIAN SPACE FORMS

1. φ-Sectional Curvature.

In section 3 of the last chapter we saw that on a K-contact manifold the sectional curvature of a plane section containing the characteristic vector ξ is equal to 1. In particular Sasakian manifolds have this property. In this section we introduce the notion of φ-sectional curvature, a notion similar to that of holomorphic sectional curvature on a Kähler manifold and give some of its properties.

Let M^{2n+1} be a Sasakian manifold with structure tensors (φ,ξ,η,g) and define a tensor field P of type $(0,4)$ by

$$P(X,Y;Z,W) = d\eta(X,Z)g(Y,W) - d\eta(X,W)g(Y,Z)$$
$$-d\eta(Y,Z)g(X,W) + d\eta(Y,W)g(X,Z).$$

Note that

$$P(X,Y;Z,W) = - P(Z,W;X,Y).$$

If $\{X,Y\}$ is an orthonormal pair orthogonal to ξ and if we set $g(X,\varphi Y) = \cos\theta$, $0 \leq \theta \leq \pi$, then

$$P(X,Y;X,\varphi Y) = - \sin^2\theta .$$

We now prove the following lemma (Moskal [43], Okumura [47]).

<u>Lemma</u>. On a Sasakian manifold we have

a) $g(R_{XY}Z,\varphi W) + g(R_{XY}\varphi Z,W) = -P(X,Y;Z,W)$. For X,Y,Z and W orthogonal to ξ we have

b) $g(R_{\varphi X\varphi Y}\varphi Z,\varphi W) = g(R_{XY}Z,W)$

and

c) $$g(R_{X\varphi X}Y,\varphi Y) = g(R_{XY}X,Y) + g(R_{X\varphi Y}X,\varphi Y) - 2P(X,Y;X,\varphi Y).$$

<u>Proof</u>. A direct computation or the Ricci identity shows that

$$(\nabla_X\nabla_Y\Phi - \nabla_Y\nabla_X\Phi - \nabla_{[X,Y]}\Phi)(Z,W) = -g(R_{XY}Z,\varphi W) - g(R_{XY}\varphi Z,W).$$

Computing the left hand side using the Sasakian condition $(\nabla_X\varphi)Y = g(X,Y)\xi - \eta(Y)X$ we obtain a). For b) using a) and the definition of P we have

$$\begin{aligned} g(R_{\varphi X\varphi Y}\varphi Z,\varphi W) &= g(R_{\varphi X\varphi Y}Z,W) - P(\varphi X,\varphi Y;Z,\varphi W) \\ &= g(R_{XY}Z,W) - P(Z,W;X,\varphi Y) \\ &\quad - P(\varphi X,\varphi Y;Z,\varphi W) \\ &= g(R_{XY}Z,W). \end{aligned}$$

Finally applying a) to the Bianchi identity

$$g(R_{X\varphi X}Y,\varphi Y) = g(R_{XY}\varphi X,\varphi Y) - g(R_{X\varphi Y}\varphi X,Y)$$

we obtain c).

A plane section in T_mM^{2n+1} is called a φ-section if there exists a vector $X \in T_mM^{2n+1}$ orthogonal to ξ such that $\{X,\varphi X\}$ is an orthonormal basis of the plane section. The sectional curvature $K(X,\varphi X)$, denoted $H(X)$, is called a φ-sectional curvature.

Recall that the sectional curvatures of a Riemannian manifold determine the curvature transformation. Similarly, it also is well known that the holomorphic sectional curvatures of a Kähler manifold determine the curvature completely. We shall show that on a Sasakian manifold the φ-sectional curvatures determine the curvature completely (Moskal [43]).

Let $B(X,Y) = g(R_{XY}Y,X)$ and for X orthogonal to ξ, $D(X) = B(X,\varphi X)$.

Proposition. On a Sasakian manifold, for tangent vectors X and Y orthogonal to ξ we have

$$B(X,Y) = \frac{1}{32}[3D(X+\varphi Y) + 3D(X-\varphi Y) - D(X+Y) - D(X-Y) - 4D(X) - 4D(Y) - 24P(X,Y;X,\varphi Y)].$$

Proof. A direct expansion gives

$$\frac{1}{32}[3D(X+\varphi Y) + 3D(X-\varphi Y) - D(X+Y) - D(X-Y) - 4D(X) - 4D(Y) - 24P(X,Y;X,\varphi Y)]$$

$$= \frac{1}{32}[6g(R_{XY}Y,X) + 6g(R_{\varphi X\varphi Y}\varphi Y,\varphi X) + 8g(R_{X\varphi X}\varphi Y,Y)$$

$$+ 12g(R_{XY}\varphi Y,\varphi X) - 2g(R_{X\varphi Y}\varphi Y,X) - 2g(R_{\varphi XY}Y,\varphi X)$$

$$+ 4g(R_{X\varphi Y}Y,\varphi X) - 24P(X,Y;X,\varphi Y)]$$

$$= g(R_{XY}Y,X)$$

using the lemma.

Proposition. Let M^{2n+1} be a Sasakian manifold and $\{X,Y\}$ an orthonormal pair in T_mM^{2n+1} with X and Y orthogonal to ξ. Set $g(X,\varphi Y) = \cos\theta$, $0 \le \theta \le \pi$. Then the sectional curvature $K(X,Y)$ is given by

$$K(X,Y) = \frac{1}{8}[3(1+\cos\theta)^2H(X+\varphi Y) + 3(1-\cos\theta)^2H(X-\varphi Y)$$

$$- H(X+Y) - H(X-Y) - H(X) - H(Y)$$

$$+ 6\sin^2\theta].$$

Proof. $K(X,Y) = B(X,Y)$ in the previous proposition, so we examine the terms in the expansion of $B(X,Y)$. Clearly $D(X) = g(X,X)^2H(X)$ for any X orthogonal to ξ, and so for the pair $\{X,Y\}$ as given $g(X+\varphi Y,X+\varphi Y) = 2(1+\cos\theta)$, $g(X-\varphi Y,X-\varphi Y) = 2(1-\cos\theta)$, $g(X+Y,X+Y) = 2$ and $g(X-Y,X-Y) = 2$. Thus $D(X+\varphi Y) = 4(1+\cos\theta)^2H(X+\varphi Y)$ and so on. As observed at the outset $P(X,Y;X,\varphi Y) = -\sin^2\theta$ completing the proof.

Theorem. The φ-sectional curvatures determine the curvature of a Sasakian manifold completely.

Proof. Since the sectional curvatures of a Riemannian manifold determine the curvature, it suffices to show that for an orthonormal pair of tangent vectors $\{X,Y\}$, $K(X,Y)$ is determined uniquely by H and g. If X and Y are orthogonal to ξ, the previous proposition applies. If X or Y is ξ, $K(X,Y) = 1$. So suppose that $X = \eta(X)\xi + aZ$ and $Y = \eta(Y)\xi + bW$ where $\eta(X), \eta(Y)$, $a = \sqrt{1 - \eta(X)^2}$ and $b = \sqrt{1 - \eta(Y)^2}$ are non-zero. Note that Z and W are unit vectors orthogonal to ξ. Recall that $R_{\xi Z}\xi = -Z$ (section IV.3) and $R_{Z\xi}Z = -\xi$ (section IV.5) for any unit vector Z orthogonal to ξ on a Sasakian manifold. Therefore

$$\begin{aligned} K(X,Y) &= g(R_{\eta(X)\xi+aZ,\eta(Y)\xi+bW}\eta(Y)\xi + bW, \eta(X)\xi + aZ) \\ &= b^2\eta(X)^2 - 2ab\,\eta(X)\,\eta(Y)g(Z,W) + a^2\eta(Y)^2 \\ &\quad + a^2b^2 g(R_{ZW}W,Z). \end{aligned}$$

Now $g(Z,W) = \frac{1}{ab}\, g(X - \eta(X)\xi,\ Y - \eta(Y)\xi) = -\frac{1}{ab}\,\eta(X)\eta(Y)$ and so $g(R_{ZW}W,Z) = [1 - g(Z,W)^2]K(Z,W) = [1 - \frac{1}{a^2b^2}\,\eta(X)^2\eta(Y)^2]K(Z,W)$. Thus

$$\begin{aligned} K(X,Y) &= \eta(X)^2(1 - \eta(Y)^2) + 2\eta(X)^2\eta(Y)^2 + \eta(Y)^2(1 - \eta(X)^2) \\ &\quad + [(1 - \eta(X)^2)(1 - \eta(Y)^2) - \eta(X)^2\eta(Y)^2]K(Z,W) \\ &= \eta(X)^2 + \eta(Y)^2 + [1 - \eta(X)^2 - \eta(Y)^2]K(Z,W) \end{aligned}$$

and $K(Z,W)$ is given by the previous proposition completing the proof.

Note that the above proof uses not only the values of the φ-sectional curvatures, but also the fact that on a Sasakian manifold

$$R_{\xi X}\xi = -X \tag{1}$$

and

$$R_{X\xi}X = -\xi \tag{2}$$

for any unit vector X orthogonal to ξ. Thus we have actually proved that any tensor field of type (1,3) on a Sasakian manifold which satisfies the symmetries of the curvature tensor, the Bianchi identity, identity a) of the Lemma, equations (1) and (2) and which agrees with the values of the φ-sectional curvatures must be the curvature tensor. We can therefore easily prove the following theorem of Ogiue [45].

<u>Theorem</u>. If the φ-sectional curvature at any point of a Sasakian manifold of dimension ≥ 5 is independent of the choice of φ-section at the point, then it is constant on the manifold and the curvature tensor is given by

$$\begin{aligned} R_{XY}Z = \frac{c+3}{4}&(g(Y,Z)X - g(X,Z)Y) \\ &+ \frac{c-1}{4}(\eta(X)\eta(Z)Y - \eta(Y)\eta(Z)X + g(X,Z)\eta(Y)\xi - g(Y,Z)\eta(X)\xi \\ &+ \Phi(Z,Y)\varphi X - \Phi(Z,X)\varphi Y + 2\Phi(X,Y)\varphi Z) \end{aligned} \tag{3}$$

where c is the constant φ-sectional curvature.

Proof. In view of the above remark in order to see that the curvature tensor has the above form with c a function on the manifold one need only check the necessary conditions and this is easily done. The Ricci tensor S and the scalar curvature ρ are given by

$$S(X,Y) = \frac{n(c+3) + c - 1}{2} g(X,Y) - \frac{(n+1)(c-1)}{2} \eta(X)\eta(Y)$$

and

$$\rho = \frac{1}{2}(n(2n+1)(c+3) + n(c-1)).$$

Now from the second Bianchi identity

$$\nabla_\alpha \rho - 2\nabla_\beta R_\alpha{}^\beta = 0$$

where $R_\alpha{}^\beta$ are the components of the Ricci tensor of type (1,1) and hence

$$(n-1)dc + (\xi c)\eta = 0.$$

Applying this to ξ, we have $\xi c = 0$ and hence $dc = 0$ for $n \neq 1$ as desired.

A Sasakian manifold of constant φ-sectional curvature c will be called a Sasakian space form and denoted $M^{2n+1}(c)$.

2. Examples of Sasakian Space Forms.

A. S^{2n+1}. We have seen that the usual contact structure induced on the unit sphere in $\mathbb{C}^{n+1}$ has the metric of constant curvature 1 as an associated metric. We denote this contact metric structure by (φ,ξ,η,g) and consider the deformed structure

$$\eta^* = \alpha\eta\,,\ \xi^* = \frac{1}{\alpha}\xi\,,\ \varphi^* = \varphi,$$
$$g^* = \alpha g + \alpha(\alpha-1)\eta \otimes \eta$$

where α is a positive constant. Such a deformation is called a D-homothetic deformation, since the metrics restricted to the contact distribution D are homothetic. It was shown by Tanno [72,74] that S^{2n+1} with this structure is a Sasakian space form with constant φ-sectional curvature $c = \frac{4}{\alpha} - 3$.

B. $\mathbb{R}^{2n+1}$. We have seen that $\mathbb{R}^{2n+1}$ with coordinates (x^i,y^i,z), $i = 1,\cdots, n$ admits a Sasakian structure

$$\eta = \frac{1}{2}(dz - \textstyle\sum_{i=1}^{n} y^i dx^i),\ g = \frac{1}{4}(\eta \otimes \eta + \textstyle\sum_{i=1}^{n}((dx^i)^2 + (dy^i)^2)).$$

With this metric $\mathbb{R}^{2n+1}$ is a Sasakian space form with $c = -3$ [48].

C. $B^n \times \mathbb{R}$. Let B^n be a simply connected bounded domain in $\mathbb{C}^n$ and (J,G) a Kähler structure with constant holomorphic sectional curvature $k < 0$. Since the fundamental

2-form Ω of the Kähler structure is closed, $\Omega = d\omega$ for some real analytic 1-form ω. Let t denote the coordinate on $\mathbb{R}$ and $\eta = \pi^*\omega + dt$. Regarding η as a connection form on the trivial line bundle, η and $g = \pi^*G + \eta \otimes \eta$ define a Sasakian structure with curvature form $d\eta = \pi^*\Omega$. By a direct computation (Ogiue [46]) $K(\tilde{\pi}X, \tilde{\pi}Y) = K_*(X,Y) - 3\eta(\nabla_{\tilde{\pi}X}\tilde{\pi}Y)^2$ where $\{X,Y\}$ is an orthonormal pair on B^n, $\tilde{\pi}$ denotes the horizontal lift and K_* the sectional curvature of B^n. Now $g(\nabla_{\tilde{\pi}X}\tilde{\pi}Y, \xi) = -g(\tilde{\pi}Y, \nabla_{\tilde{\pi}X}\xi) = g(\tilde{\pi}Y, \varphi\tilde{\pi}X)$; thus since $\varphi\tilde{\pi}X = \tilde{\pi}JX$, $B^n \times \mathbb{R}$ has constant φ-sectional curvature $c = k - 3$ (Tanno [74]).

3. Integral Submanifolds of Sasakian Space Forms.

We have already seen that the maximum dimension of an integral submanifold of the contact distribution of a contact manifold M^{2n+1} is n. In example III. 2.B we saw that the 2-dimensional torus T^2 is an integral surface of the contact distribution on S^5 which is minimally immersed but not totally geodesic. Example III. 2. A gave S^2 as a totally geodesic integral surface of S^5. In this section we will study n-dimensional integral submanifolds minimally immersed in a Sasakian space form $M^{2n+1}(c)$.

Let $M^{2n+1}(c)$ be a Sasakian space form with structure tensors (φ, ξ, η, G) and let $\tilde{\nabla}$ denote the Riemannian connection of G. Now consider an integral submanifold $\iota: M^n \to M^{2n+1}(c)$. The induced metric g is given by $g(X,Y) \circ \iota = G(\iota_* X, \iota_* Y)$; for brevity, however, we shall not distinguish notationally between X and $\iota_* X$. Let $X_1, \cdots, X_n$ be a local orthonormal basis of vector fields on M^n. Then by the first proposition of section III.1 $\xi_0 = \xi$, $\xi_i = \varphi X_i$, $i = 1, \cdots, n$ form a local field of orthonormal normal vectors. Let ∇ denote the Riemannian connection of g and $\nabla^{\perp}$ the connection in the normal bundle. Then the Gauss-Weingarten equations are

$$\tilde{\nabla}_X Y = \nabla_X Y + \sigma(X,Y)$$

$$\tilde{\nabla}_X \xi_\alpha = - A_\alpha X + \nabla^{\perp}_X \xi_\alpha, \quad \alpha = 0, 1, \cdots, n$$

where σ is the second fundamental form and the A_α's the

Weingarten maps. Decomposing σ we have $\sigma(X,Y) = \sum_\alpha h^\alpha(X,Y)$ where the tensors h^α satisfy $h^\alpha(X,Y) = g(A_\alpha X,Y)$ and are symmetric. Letting R and $\tilde{R}$ denote the curvature transformations of ∇ and $\tilde{\nabla}$ respectively, the equation of Gauss is

$$g(R_{XY}Z,W) = G(\tilde{R}_{XY}Z,W) + G(\sigma(X,W),\sigma(Y,Z))$$
$$- G(\sigma(X,Z),\sigma(Y,W)).$$

Finally for the second fundamental form σ, we define the covariant derivative $'\nabla$ with respect to the connection in the (tangent bundle) ⊕ (normal bundle) by

$$('\nabla_X\sigma)(Y,Z) = \nabla_X^\perp(\sigma(Y,Z)) - \sigma(\nabla_XY,Z) - \sigma(Y,\nabla_XZ).$$

<u>Lemma 1</u>. $h^0 = 0$.

<u>Proof</u>. Note that for this lemma we need only assume that the ambient manifold is a K-contact manifold; the proof is simply:

$$h^0(X,Y) = G(\tilde{\nabla}_XY,\xi) = -G(Y,\tilde{\nabla}_X\xi) = G(Y,\varphi X) = 0.$$

Let $\omega^1,\cdots,\omega^n,\omega^{1^*},\cdots,\omega^{n^*}$, $\omega^0 = \eta$ be the dual basis of $X_i,\varphi X_i$, $\xi, i = 1,\cdots,n$. Then the first structural equation of Cartan is

$$d\omega^A = -\sum_{B=0}^{2n} \omega^A_B \wedge \omega^B,\ n+i=i^*$$

where (ω^A_B) is a real representation of a skew-Hermitian matrix and hence $\omega^{i^*}_j = \omega^{j^*}_i$. Now as ω^α and $d\omega^\alpha$ vanish along M^n, $\sum_B \omega^\alpha_B \wedge \omega^B = 0$ in which the ω^α_i's give the second fundamental form, i.e.

$$\omega^{i^*}_j = \sum_k h^i_{jk}\omega^k , \quad \omega^0_i = \sum_j h^0_{ij}\omega^j = 0$$

where $h^\alpha_{ij} = h^\alpha(X_i, X_j)$. We now obtain the following algebraic lemma.

<u>Lemma 2</u>. Let M^n be an immersed submanifold of an almost contact manifold M^{2n+1} such that the condition of the first proposition of section III.1 holds. Then the Weingarten maps A_i, $i = 1, \cdots, n$ satisfy

1) $A_i X_j = A_j X_i$,

2) $\operatorname{tr}(\sum_i A_i^2)^2 = \sum_{i,j} (\operatorname{tr} A_i A_j)^2$.

<u>Proof</u>. We noted above that $\omega^{i^*}_j = \omega^{j^*}_i$ from which $h^i_{jk} = h^j_{ik}$, but $h^\alpha_{jk} = g(A_\alpha X_j, X_k)$ giving 1). For 2) we have

$$\operatorname{tr}(\sum_i A_i^2)^2 = \sum_{i,j} \operatorname{tr} A_i^2 A_j^2 = \sum h^i_{k\ell} h^i_{\ell m} h^j_{mn} h^j_{nk}$$

$$= \sum h^k_{i\ell} h^m_{\ell i} h^m_{jn} h^k_{nj} = \sum_{k,m} (\operatorname{tr} A_k A_m)^2$$

where the sums are over all repeated indices.

For M^n as an integral submanifold of a Sasakian space form, the Gauss equation and the special form of the curvature transformation $\tilde{R}$ (equation (3) of section 1) give

$$g(R_{XY}Z,W) = \frac{c+3}{4}(g(X,W)g(Y,Z) - g(X,Z)g(Y,W)) + \sum_\alpha (g(A_\alpha X,W)g(A_\alpha Y,Z) - g(A_\alpha X,Z)g(A_\alpha Y,W)) \tag{4}$$

and hence the sectional curvature $K(X,Y)$ of M^n determined by an orthonormal pair X,Y is given by

$$K(X,Y) = \frac{c+3}{4} + \sum_\alpha (g(A_\alpha X,X)g(A_\alpha Y,Y) - g(A_\alpha X,Y)^2).$$

Moreover the Ricci tensor S and the scalar curvature ρ of M^n are given by

$$S(X,Y) = \frac{n-1}{4}(c+3)g(X,Y) + \sum_\alpha (\mathrm{tr}\, A_\alpha) g(A_\alpha X,Y) - \sum_\alpha g(A_\alpha X, A_\alpha Y),$$

$$\rho = \frac{n(n-1)}{4}(c+3) + \sum_\alpha (\mathrm{tr}\, A_\alpha)^2 - \|\sigma\|^2$$

where $\|\sigma\|^2 = \sum_\alpha \mathrm{tr}(A_\alpha^2) = \sum_{\alpha,i,j} h^\alpha_{ij} h^\alpha_{ij}$ is the square of the length of the second fundamental form. In particular, if the immersion is minimal,

$$S(X,Y) = \frac{n-1}{4}(c+3)g(X,Y) - \sum_\alpha g(A_\alpha X, A_\alpha Y), \tag{5}$$

$$\rho = \frac{n(n-1)}{4}(c+3) - \|\sigma\|^2 \tag{6}$$

Proposition. Let M^n be an integral submanifold of a Sasakian space form $M^{2n+1}(c)$ which is minimally immersed. Then the following are equivalent.

a) M^n is totally geodesic

b) M^n is of constant curvature $k = \frac{c+3}{4}$

c) $S = \frac{n-1}{4}(c+3)g$

d) $\rho = \frac{n(n-1)}{4}(c+3)$

Proof. That a) implies b), c) and d) is immediate from the above equations. That c) and d) each imply a) is also immediate. For b) $\Rightarrow$ a), let X_1 be an arbitrary unit vector and choose $X_2, \cdots, X_n$ such that $X_1, X_2, \cdots, X_n$ is an orthonormal basis. Then

$$S(X_1,X_1) = \sum_{i=2}^{n} K(X_1,X_i) = \frac{c+3}{4}(n-1)$$

which is c).

In [16], Chern, do Carmo and Kobayashi obtained a formula for the Laplacian of the square of the length of the second fundamental form of a minimal immersion into a locally symmetric space. Taking into account the special form of the curvature tensor of $M^{2n+1}(c)$ the same formula holds for an integral submanifold M^n even though $M^{2n+1}(c)$ is not locally symmetric, namely

$$\begin{aligned} \frac{1}{2}\Delta\|\sigma\|^2 = \|'\nabla\sigma\|^2 + \sum_{\alpha,\beta} \operatorname{tr}(A_\alpha A_\beta - A_\beta A_\alpha)^2 - \sum_{\alpha,\beta} (\operatorname{tr} A_\alpha A_\beta)^2 \\ - \Sigma\,(4\tilde{R}^{\alpha}{}_{\beta ij}h^{\alpha}_{jk}h^{\beta}_{ik} - \tilde{R}^{\alpha}{}_{k\beta k}h^{\alpha}_{ij}h^{\beta}_{ij} \\ + 2\tilde{R}^{i}{}_{jkj}h^{\alpha}_{i\ell}h^{\alpha}_{k\ell} + 2\tilde{R}^{i}{}_{jk\ell}h^{\alpha}_{i\ell}h^{\alpha}_{jk}) \end{aligned} \tag{7}$$

where the $\tilde{R}^A{}_{BCD}$'s are the components of the curvature tensor of $\tilde{\nabla}$ and the sum is over all repeated indices. Using equation (3) of section 1 for the curvature tensor of $M^{2n+1}(c)$ the last term on the right of equation (7) becomes $\frac{n(c+3)+c-1}{4}\|\sigma\|^2$. Using Lemma 2 we then obtain

$$\frac{1}{2}\Delta\|\sigma\|^2 = \|'\nabla\sigma\|^2 + \sum_{\alpha,\beta} \operatorname{tr}(A_\alpha A_\beta)^2 - 3\sum_{\alpha,\beta}(\operatorname{tr} A_\alpha A_\beta)^2 + \frac{n(c+3)+c-1}{4}\|\sigma\|^2 . \tag{8}$$

The equation of Gauss and Lemma 2 also give

$$\sum (h^\alpha_{ij}h^\alpha_{k\ell}R_{\ell ijk} + h^\alpha_{ij}h^\alpha_{i\ell}R_{\ell kjk}) = \frac{n(c+3)}{4}\|\sigma\|^2 + \frac{1}{2}\sum_{\alpha,\beta}\operatorname{tr}(A_\alpha A_\beta - A_\beta A_\alpha)^2 - \sum_{\alpha,\beta}(\operatorname{tr} A_\alpha A_\beta)^2 \tag{9}$$

so we can write

$$\frac{1}{2}\Delta\|\sigma\|^2 = \|'\nabla\sigma\|^2 + \sum (h^\alpha_{ij}h^\alpha_{k\ell}R_{\ell ijk} + h^\alpha_{ij}h^\alpha_{i\ell}R_{\ell kjk}) + \frac{1}{2}\sum_{\alpha,\beta}\operatorname{tr}(A_\alpha A_\beta - A_\beta A_\alpha)^2 + \frac{c-1}{4}\|\sigma\|^2 . \tag{10}$$

Lemma 3. If the sectional curvature of M^n is greater than δ, then

$$\frac{1}{2}\Delta\|\sigma\|^2 \geq \|'\nabla\sigma\|^2 + (1+a)n\delta\|\sigma\|^2 - \frac{na(c+3)-(c-1)}{4}\|\sigma\|^2 + \frac{1-a}{2}\sum_{\alpha,\beta}\operatorname{tr}(A_\alpha A_\beta - A_\beta A_\alpha)^2 + a\sum_{\alpha,\beta}(\operatorname{tr} A_\alpha A_\beta)^2$$

for any real number $a \geq -1$.

Proof. Let $\lambda_1^\alpha, \cdots, \lambda_n^\alpha$ be the eigenvalues of A_α and suppose that the X_i's form an orthonormal basis of eigenvectors. Then

$$\sum_{i,j,k,\ell} (h_{ij}^\alpha h_{k\ell}^\alpha R_{\ell ijk} + h_{ij}^\alpha h_{i\ell}^\alpha R_{\ell kjk})$$

$$= \sum_{i,k} (\lambda_i^\alpha \lambda_k^\alpha R_{kiik} + (\lambda_i^\alpha)^2 R_{ikik})$$

$$= \frac{1}{2} \sum_{i,k} (\lambda_i^\alpha - \lambda_k^\alpha)^2 R_{ikik}$$

$$\geq \frac{1}{2} \sum_{i,k} (\lambda_i^\alpha - \lambda_k^\alpha)^2 \delta$$

$$= n\,\delta \textstyle\sum_i (\lambda_i^\alpha)^2$$

$$= n\,\delta\, \mathrm{tr}(A_\alpha^2)$$

where we have used the fact that $\sum_i \lambda_i^\alpha = 0$, i.e. the minimality. Thus we have

$$\sum_{i,j,k,\ell,\alpha} (h_{ij}^\alpha h_{k\ell}^\alpha R_{\ell ijk} + h_{ij}^\alpha h_{i\ell}^\alpha R_{\ell kjk}) \geq n\delta \|\sigma\|^2 . \qquad (11)$$

Now putting together equations (9), (10) and (11) we obtain the lemma.

Lemma 4. $\|{}'\nabla\sigma\|^2 \geq \|\sigma\|^2$.

Proof. If we denote by h_{ijk}^α the components of ${}'\nabla\sigma$, that is $h_{ijk}^\alpha = G(({}'\nabla_{X_k}\sigma)(X_i, X_j), \xi_\alpha)$ we then have

$$h^0_{ijk} = G(('\nabla_{X_k}\sigma)(X_i,X_j),\xi)$$

$$= G(\nabla^{\perp}_{X_k}(\sigma(X_i,X_j)),\xi)$$

$$= -G(\sigma(X_i,X_j),\tilde{\nabla}_{X_k}\xi)$$

$$= G(\sigma(X_i,X_j),\varphi X_k)$$

$$= h^k_{ij} .$$

Therefore

$$\|'\nabla\sigma\|^2 = \sum h^{\alpha}_{ijk}h^{\alpha}_{ijk} = \sum_{i,j,k,m=1}^{n} (h^m_{ijk})^2 + \|\sigma\|^2 \geq \|\sigma\|^2 .$$

We now prove the following theorem on minimally immersed integral submanifolds (Blair and Ogiue [10]).

<u>Theorem</u>. Let M^n be a compact integral submanifold of the contact distribution of $M^{2n+1}(c)$ which is minimally immersed. If the sectional curvature of M^n is greater than $\frac{(n-2)(c+3)}{4(2n-1)}$, then M^n is totally geodesic.

<u>Proof</u>. Let $\Lambda = (\mathrm{tr}\, A_iA_j)$. Then Λ is a symmetric $n \times n$ matrix defined with respect to an orthonormal basis which is covariant for an orthogonal change of basis. Thus, without loss of generality, we may assume that $\mathrm{tr}\, A_iA_j = 0$ for $i \neq j$.

We also have the algebraic lemma

$$\sum_{i,j} \operatorname{tr}(A_iA_j - A_jA_i)^2 \geq -2 \sum_{i\neq j} \operatorname{tr} A_i^2 \operatorname{tr} A_j^2$$

due to Chern, do Carmo and Kobayashi [16]. Thus

$$\sum_{i,j} \operatorname{tr}(A_iA_j - A_jA_i)^2 \geq -2\|\sigma\|^4 + 2\sum_i (\operatorname{tr} A_i^2)^2 .$$

Lemmas 3 and 4 now give

$$\frac{1}{2}\Delta\|\sigma\|^2 \geq (1+a)n\delta\|\sigma\|^2 - \frac{(na-1)(c+3)}{4}\|\sigma\|^2$$
$$- (1-a)\|\sigma\|^4 + \sum_i (\operatorname{tr} A_i^2)^2$$

for $-1 \leq a \leq 1$. In addition one can easily show that $\sum_{\alpha,\beta} (\operatorname{tr} A_\alpha A_\beta)^2 \geq \frac{1}{n}\|\sigma\|^4$. Thus we have

$$\frac{1}{2}\Delta\|\sigma\|^2 \geq (1+a)n\delta\|\sigma\|^2 - \frac{(na-1)(c+3)}{4}\|\sigma\|^2 + (\frac{1}{n} - (1-a))\|\sigma\|^4 .$$

Setting $a = 1 - \frac{1}{n}$, we obtain

$$\frac{1}{2}\Delta\|\sigma\|^2 \geq ((2n-1)\delta - \frac{(n-2)(c+3)}{4})\|\sigma\|^2 \tag{12}$$

which is nonnegative under the assumption of the theorem. Finally since M^n is compact, $\int_{M^n} \Delta\|\sigma\|^2 * 1 = 0$, so that

so that $\Delta\|\sigma\|^2 = 0$ and hence $\|\sigma\| = 0$ by the inequality (12).

We remark that for $n = 2$, the hypothesis on the curvature is simply that it be positive. The example

(III.2.B) gives T^2 as a flat minimally immersed non-totally geodesic integral surface of S^5; thus in the case $n=2$, the number in the theorem is best possible.

The following similar theorem was obtained by S. Yamaguchi, M. Kon and T. Ikawa [80] (see also [9,79]).

<u>Theorem</u>. Let M^n be a compact integral submanifold minimally immersed in $M^{2n+1}(c)$. If

$$\|\sigma\|^2 < \frac{n(n+1)(c+3)}{4(2n-1)}$$

or equivalently

$$\rho > \frac{n^2(n-2)(c+3)}{2(2n-1)}$$

then M^n is totally geodesic.

We now prove the following result, also due to Yamaguchi, Kon and Ikawa [80].

<u>Theorem</u>. Let M^n be a minimal integral submanifold of a Sasakian space form $M^{2n+1}(c)$. If M^n has constant curvature k, then either $k = \frac{c+3}{4}$, in which case M^n is totally geodesic, or $k \leq 0$.

<u>Proof</u>. Since M^n has constant curvature k, $\rho = n(n-1)k$ and hence equation (6) gives $\|\sigma\|^2 = n(n-1)(\frac{c+3}{4} - k)$ and so $k \leq \frac{c+3}{4}$. Also equation (4) becomes

$$\sum_\alpha (h^\alpha_{ik}h^\alpha_{j\ell} - h^\alpha_{i\ell}h^\alpha_{jk}) = (k - \frac{c+3}{4})(\delta_{ik}\delta_{j\ell} - \delta_{i\ell}\delta_{jk}) ;$$

multiplying this equation by $\sum_m h^m_{i\ell} h^m_{jk}$ and summing on repeated indices we have

$$\sum_{h,m} \operatorname{tr}(A_h A_m)^2 - \sum_{h,m} (\operatorname{tr} A_h A_m)^2 = (k - \frac{c+3}{4}) \|\sigma\|^2 .$$

Moreover $S = \frac{\rho}{n} g$ and hence equation (5) and Lemma 2 give

$$\operatorname{tr} A_j A_\ell = \sum_{i,k} h^j_{ik} h^\ell_{ki} = \sum_{i,k} h^i_{jk} h^i_{k\ell} = (\frac{n-1}{4}(c+3) - \frac{\rho}{n}) \delta_{j\ell} = \frac{\|\sigma\|^2}{n} \delta_{j\ell} .$$

Substituting these into equation (8) we have

$$0 = \|{}'\nabla\sigma\|^2 + 2(k - \frac{c+3}{4}) \|\sigma\|^2 - \frac{\|\sigma\|^4}{n} + \frac{n(c+3) + c-1}{4} \|\sigma\|^2 .$$

Using $\|\sigma\|^2 = n(n-1)(\frac{c+3}{4} - k)$ and Lemma 4,

$$\|{}'\nabla\sigma\|^2 = n(n^2-1)(k - \frac{c+3}{4})(k - \frac{1}{n+1}) \geq n(n-1)(\frac{c+3}{4} - k)$$

from which the result follows.

Let us now consider briefly the case of an integral surface of the contact distribution on the unit sphere S^5. In particular we prove the following theorem (Yamaguchi, Kon, Miyahara [81]).

<u>Theorem</u>. Let M^2 be a complete integral surface of the contact distribution on S^5 which is minimally immersed. If the Gaussian curvature K of M^2 is ≤ 0, then M^2 is flat.

<u>Proof</u>. Choose a system of isothermal coordinates (x^1, x^2) so that the induced metric g is given by $g = E((dx^1)^2 + (dx^2)^2)$. Let $X_i = \partial/\partial x^i$ and $Y_i = E^{-1/2} X_i$; then

$\{Y_1, Y_2\}$ is an orthonormal pair and $\xi_0 = \xi, \xi_i = \varphi Y_i$ are orthonormal normal vector fields. Also let $\sigma_{ij} = \sigma(X_i, X_j)$ Finally recall the standard formulas for the induced connection ∇, $\nabla_{X_1} X_1 = \frac{X_1 E}{2E} X_1 - \frac{X_2 E}{2E} X_2$, $\nabla_{X_1} X_2 = \frac{X_2 E}{2E} X_1 + \frac{X_1 E}{2E} X_2$, $\nabla_{X_2} X_2 = -\frac{X_1 E}{2E} X_1 + \frac{X_2 E}{2E} X_2$.

From the Codazzi equation we have

$$\begin{aligned} 0 = (\tilde{R}_{X_1 X_2} X_1)^{\perp} &= \nabla^{\perp}_{X_1} \sigma(X_2, X_1) - \sigma(\nabla_{X_1} X_2, X_1) - \sigma(X_2, \nabla_{X_1} X_1) \\ &\quad - \nabla^{\perp}_{X_2} \sigma(X_1, X_1) + \sigma(\nabla_{X_2} X_1, X_1) + \sigma(X_1, \nabla_{X_2} X_1) \end{aligned}$$

and therefore

$$0 = \nabla^{\perp}_{X_1} \sigma_{12} - \nabla^{\perp}_{X_2} \sigma_{11} .$$

Similarly

$$0 = \nabla^{\perp}_{X_1} \sigma_{11} + \nabla^{\perp}_{X_2} \sigma_{12} .$$

Now define a function F by

$$F = G(\sigma_{11}, \varphi X_1) - \sqrt{-1}\, G(\sigma_{12}, \varphi X_1) .$$

Note that F is nowhere zero on M^2, for if $F = 0$ at some point m then h^1_{11} and h^1_{12} vanish at m, but by minimality $h^1_{22} = -h^1_{11} = 0$ and in turn by Lemma 2, $h^2_{12} = h^1_{22} = 0$ and $-h^2_{22} = h^2_{11} = h^1_{12} = 0$. Thus σ vanish at m and so by the Gauss equation, the Gaussian curvature at m is 1 contradicting the hypothesis $K \leq 0$.

Differentiating the real and imaginary parts of F we have

$$X_1 \operatorname{Re} F = G(\nabla^{\perp}_{X_1}\sigma_{11}, \varphi X_1) + G(\sigma_{11}, E\xi + \frac{X_1 E}{2E}\varphi X_1 - \frac{X_2 E}{2E}\varphi X_2)$$

$$= -G(\nabla^{\perp}_{X_2}\sigma_{12}, \varphi X_1) + \frac{1}{2}E^{1/2}((X_1 E)h^1(Y_1,Y_1) - (X_2 E)h^2(Y_1,Y_1))$$

and

$$X_2 \operatorname{Im} F = -G(\nabla^{\perp}_{X_2}\sigma_{12}, \varphi X_1) - G(\sigma_{12}, \frac{X_2 E}{2E}\varphi X_1 + \frac{X_1 E}{2E}\varphi X_2)$$

$$= -G(\nabla^{\perp}_{X_2}\sigma_{12}, \varphi X_1) - \frac{1}{2}E^{1/2}((X_2 E)h^1(Y_1,Y_2) + (X_1 E)h^2(Y_1,Y_2))$$

and hence $X_1 \operatorname{Re} F = X_2 \operatorname{Im} F$. In like manner we obtain $X_2 \operatorname{Re} F = -X_1 \operatorname{Im} F$. Thus F is analytic and therefore $\log |F|^2$ is harmonic.

Now $|F|^2 = G(\sigma_{11}, \varphi X_1)^2 + G(\sigma_{12}, \varphi X_1)^2 = E^3(h^1(Y_1,Y_1)^2 + h^1(Y_1,Y_2)^2)$. On the other hand the Gauss equation gives the Gaussian curvature K as

$$K = 1 + G(\sigma(Y_1,Y_1), \sigma(Y_2,Y_2)) - G(\sigma(Y_1,Y_2), \sigma(Y_1,Y_2))$$

$$= 1 - h^1(Y_1,Y_1)^2 - h^2(Y_1,Y_1)^2 - h^1(Y_1,Y_2)^2 - h^2(Y_1,Y_2)^2$$

$$= 1 - 2(h^1(Y_1,Y_1)^2 + h^1(Y_1,Y_2)^2).$$

Thus we have $|F|^2 = E^3 \frac{1-K}{2}$. Note also the classical formula for the Gaussian curvature of g, namely $K = \frac{-1}{6E}\Delta \log E^3$.

Suppose now that the Gaussian curvature of M^2 is non-positive. Then

$$\Delta \log \frac{|F|^2}{E^3} = - \Delta \log E^3 = 6EK \leq 0 \qquad (13)$$

and

$$\log \frac{|F|^2}{E^3} = \log \frac{1-K}{2} \geq \log \frac{1}{2} .$$

Thus $-\log \frac{|F|^2}{E^3}$ is a subharmonic function which is bounded above.

Now define a metric g^* on M^2 by

$$g^* = |F|((dx^1)^2 + (dx^2)^2) ;$$

its Gaussian curvature is $-\frac{1}{4|F|} \Delta \log |F|^2 = 0$. That is, g^* is a flat metric on M^2 which is conformally equivalent to g and hence the universal covering surface $\tilde{M}$ of M^2 is conformally equivalent to the Euclidean plane. Thus $\tilde{M}$ is a parabolic surface; but every subharmonic function which is bounded above on a parabolic surface is a constant. Therefore $-\log \frac{|F|^2}{E^3}$, lifted to $\tilde{M}$, is a constant and hence it is a constant function on M^2. Equation (13) now gives $K = 0$.

Combining this theorem with the preceding one we have the following corollary.

<u>Corollary</u>. A complete integral surface of S^5 with constant curvature which is minimally immersed has constant curvature 0 or 1.

CHAPTER VI

NON-EXISTENCE OF FLAT CONTACT METRIC STRUCTURES

1. Non-Existence of Flat Contact Metric Structures in Dimension ≥ 5.

We have seen that for a K-contact manifold the sectional curvatures of plane sections containing the characteristic vector ξ are equal to 1. In this section we shall show that every contact metric manifold of dimension ≥ 5 must have some curvature though not necessarily in the plane sections containing ξ [6]. In dimension 3 flat associated metrics do exist as we shall see in section 2.

Theorem. Let M^{2n+1} be a contact manifold of dimension ≥ 5. Then M^{2n+1} cannot admit a contact metric structure of vanishing curvature.

Proof. The proof will be by contradiction. We let (φ,ξ,η,g) denote the structure tensors of a contact metric structure and suppose that g is flat. In section IV.3 (equation (11)) we showed that for a contact metric structure

$$\frac{1}{2}(R_{\xi X}\xi - \varphi R_{\xi\varphi X}\xi) = h^2X + \varphi^2X$$

where $h = \frac{1}{2}\mathcal{L}_\xi\varphi$. Thus if g is flat $h^2 = -\varphi^2$; and hence $h\xi = 0$ and rank $(h) = 2n$. The eigenvectors corresponding to the non-zero eigenvalues of h are orthogonal to ξ and the non-zero eigenvalues are ± 1. Recall that $d\eta(X,Y)$

$= \frac{1}{2}(g(\nabla_X\xi, Y) - g(\nabla_Y\xi, X))$ and that for a contact metric structure

$$\nabla_X\xi = -\varphi hX - \varphi X \tag{1}$$

(section IV.3, equation (10)). Thus

$$2g(X,\varphi Y) = g(-\varphi hX - \varphi X, Y) - g(-\varphi hY - \varphi Y, X)$$

giving

$$g(\varphi hX, Y) = g(\varphi hY, X) = -g(h\varphi X, Y),$$

that is h and φ anti-commute. In particular, if X is an eigenvector of the eigenvalue $+1$, φX is an eigenvector of -1 and vice-versa. Thus the contact distribution D defined by $\eta = 0$ is decomposed into the orthogonal eigenspaces of ± 1 which we denote by $[+1]$ and $[-1]$.

We now show that the distribution $[-1]$ is integrable. If X and Y are vector fields belonging to $[-1]$, equation (1) gives $\nabla_X\xi = 0$ and $\nabla_Y\xi = 0$. Thus since M^{2n+1} is flat

$$0 = R_{XY}\xi = -\nabla_{[X,Y]}\xi = \varphi h[X,Y] + \varphi[X,Y];$$

but $\eta([X,Y]) = -2d\eta(X,Y) = -2g(X,\varphi Y) = 0$, so that $h[X,Y] = -[X,Y]$. Applying the same argument to ξ and $X \in [-1]$ we see that the distribution $[-1] \oplus [\xi]$ spanned by $[-1]$ and ξ is also integrable.

Since $[-1] \oplus [\xi]$ is integrable, we can choose local coordinates $(u^0, \cdots, u^{2n})$ such that $\partial/\partial u^0, \cdots, \partial/\partial u^n \in [-1] \oplus [\xi]$ and we define local vector fields X_i, $i = 1, \cdots, n$ by $X_i = \partial/\partial u^{n+i} + \sum_{j=0}^{n} f_i^j \partial/\partial u^j$ where the f_i^j's are functions chosen so that $X_i \in [+1]$. Thus $X_1, \cdots, X_n$ are n linearly independent vector fields spanning $[+1]$. Clearly $[\partial/\partial u^k, X_i] \in [-1] \oplus [\xi]$ for $k = 0, \cdots, n$ and hence ξ is parallel along $[\partial/\partial u^k, X_i]$. Therefore using equation (1) and the vanishing curvature

$$0 = \nabla_{[\partial/\partial u^k, X_i]} \xi = \nabla_{\partial/\partial u^k} \nabla_{X_i} \xi - \nabla_{X_i} \nabla_{\partial/\partial u^k} \xi = -2 \nabla_{\partial/\partial u^k} \varphi X_i$$

from which we have

$$\nabla_{\varphi X_j} \varphi X_i = 0 . \tag{2}$$

Similarly, noting that $[X_i, X_j] \in [-1]$,

$$0 = R_{X_i X_j} \xi = -2 \nabla_{X_i} \varphi X_j + 2 \nabla_{X_j} \varphi X_i$$

giving

$$\nabla_{X_i} \varphi X_j = \nabla_{X_j} \varphi X_i \tag{3}$$

or equivalently

$$\varphi [X_i, X_j] = -(\nabla_{X_i} \varphi) X_j + (\nabla_{X_j} \varphi) X_i . \tag{4}$$

Using equations (2) and (1)

$$0 = R_{X_i \varphi X_j}\xi = -\nabla_{[X_i, \varphi X_j]}\xi = \varphi h[X_i, \varphi X_j] + \varphi[X_i, \varphi X_j]$$

from which

$$g([X_i, \varphi X_j], X_k) = -g(h[X_i, \varphi X_j], X_k) = -g([X_i, \varphi X_j], X_k)$$

and hence

$$g([X_i, \varphi X_j], X_k) = 0 . \qquad (5$$

We now compute $(\nabla_{X_i}\varphi)X_j$ explicitly. Using the general formula for the covariant derivative of φ (section IV.1, equation (7)) and equations (2), (5) and (4) above we have

$$2g((\nabla_{X_i}\varphi)X_j, X_k) = g([\varphi,\varphi](X_j, X_k), \varphi X_i)$$

$$= -g([X_j, X_k], \varphi X_i)$$

$$= g(-(\nabla_{X_j}\varphi)X_k + (\nabla_{X_k}\varphi)X_j, X_i) .$$

Since $\Phi = d\eta$, the sum of the cyclic permutations of i, j and k in $g((\nabla_{X_i}\varphi)X_j, X_k)$ is zero. Thus our computation yields $g((\nabla_{X_i}\varphi)X_j, X_k) = 0$. Similarly

$$2g((\nabla_{X_i}\varphi)X_j, \varphi X_k) = g([\varphi,\varphi](X_j, \varphi X_k), \varphi X_i)$$

$$= g(-[X_j, \varphi X_k] - [\varphi X_j, X_k], \varphi X_i)$$

$$= g(-\nabla_{X_j}\varphi X_k + \nabla_{\varphi X_k}X_j - \nabla_{\varphi X_j}X_k + \nabla_{X_k}\varphi X_j, \varphi X_i)$$

which vanishes by equations (2) and (3). Finally

$$2g((\nabla_{X_i}\varphi)X_j,\xi) = g(\varphi^2[X_j,\xi],\varphi X_i) + 2d\eta(\varphi X_j,X_i)$$

$$= 4g(X_j,X_i).$$

Thus for any vector fields X and Y in $[+1]$ on a contact metric manifold such that ξ is annihilated by the curvature transformation,

$$(\nabla_X\varphi)Y = 2g(X,Y)\xi. \tag{6}$$

Note that equation (4) now gives $[X_i,X_j] = 0$.

Before differentiating equation (6) we show that $\nabla_{X_i}X_j \in [+1]$. First note that

$$-2g(\nabla_{\varphi X_i}X_j,X_k) = 2g((\nabla_{\varphi X_i}\varphi)X_j,\varphi X_k) = 0 \tag{7}$$

by a computation of the type we have been doing. Therefore

$$g(\nabla_{X_i}X_j,\varphi X_k) = -g(X_j,\nabla_{X_i}\varphi X_k) = -g(X_j,[X_i,\varphi X_k]) = 0$$

by equation (5). That $g(\nabla_{X_i}X_j,\xi) = 0$ is trivial and so $\nabla_{X_i}X_j \in [+1]$.

Now to show the non-existence of flat contact metric structures for contact manifolds of dimension greater than or equal to 5, we will contradict the linear independence of the X_i's. Equation (6) can be written as

$$\nabla_{X_i}\varphi X_j - \varphi\nabla_{X_i}X_j = 2g(X_i,X_j)\xi,$$

Differentiating this we have

$$\nabla_{X_k}\nabla_{X_i}\varphi X_j - (\nabla_{X_k}\varphi)\nabla_{X_i}X_j - \varphi\nabla_{X_k}\nabla_{X_i}X_j$$

$$= 2(X_k g(X_i,X_j))\xi - 4g(X_i,X_j)\varphi X_k .$$

Taking the inner product with φX_ℓ, remembering equation (6) and that $\nabla_{X_i}X_j \in [+1]$, we have

$$g(\nabla_{X_k}\nabla_{X_i}\varphi X_j,\varphi X_\ell) - g(\nabla_{X_k}\nabla_{X_i}X_j,X_\ell) = -4g(X_i,X_j)g(X_k,X_\ell).$$

Interchanging i and k, $i \neq k$ and subtracting we have

$$0 = g(X_i,X_j)g(X_k,X_\ell) - g(X_k,X_j)g(X_i,X_\ell)$$

by virtue of the flatness and $[X_i,X_k] = 0$. Setting $i=j$ and $k=\ell$ we have $0 = g(X_i,X_i)g(X_k,X_k) - g(X_i,X_k)^2$ contradicting the linear independence of X_i and X_k.

Note that in the proof of our theorem, the vanishing of $R_{\xi X}\xi$ is enough to obtain the decomposition of the contact distribution into ± 1 eigenspaces of the operator $h = \frac{1}{2}\mathcal{L}_\xi\varphi$. Moreover $R_{XY}\xi = 0$ for X and Y in $[-1]$ is sufficient for the integrability of $[-1]$. Thus we have the following result [6].

<u>Theorem</u>. Let M^{2n+1} be a contact manifold with contact metric structure (φ,ξ,η,g). If the sectional curvatures of all plane sections containg ξ vanish, then the operator h has rank $2n$ and the contact distribution is decomposed into the ± 1 eigenspaces of h. Moreover if $R_{XY}\xi = 0$

for $X,Y \in [-1]$, M admits a foliation by n-dimensional integral submanifolds of the contact distribution along which ξ is parallel.

In the next section we shall se that $\mathbb{R}^{2n+1}$ with the usual contact form carries an associated metric satisfying the conditions of this theorem.

If one assumes even more, namely that the characteristic vector field ξ is annihilated by the curvature transformation, we obtain the following result [7] and in the next chapter we shall see that the tangent sphere bundle of a flat Riemannian manifold admits such a structure as a contact metric manifold.

<u>Theorem</u>. Let M^{2n+1} be a contact metric manifold and suppose that $R_{XY}\xi = 0$ for all vector fields X and Y. Then M^{2n+1} is locally the product of a flat (n+1)-dimensional manifold and an n-dimensional manifold of positive constant curvature equal to 4.

<u>Proof</u>. We noted in the above proof that $[X_i,X_j] = 0$ so that the distribution $[+1]$ is also integrable and hence we may take $X_i = \partial/\partial u^{n+i}$. Moreover M^{2n+1} is locally the product of an integral submanifold M^{n+1} of $[-1] \oplus [\xi]$ and an integral submanifold M^n of $[+1]$. Since $\{\varphi X_i, \xi\}$ is a local basis of tangent vector fields on M^{n+1}, equation (2) and $R_{XY}\xi = 0$ show that M^{n+1} is flat.

Now $\nabla_{\varphi X_i} X_j = 0$, for by equation (7) $g(\nabla_{\varphi X_i} X_j, X_k) = 0$ by equation (2) $g(\nabla_{\varphi X_i} X_j, \varphi X_k) = 0$, and $g(\nabla_{\varphi X_i} X_j, \xi) = 0$ is trivial. Interchanging i and k in equation (8) and subtracting we have

$$g(R_{X_k X_i} \varphi X_j, \varphi X_\ell) - g(R_{X_k X_i} X_j, X_\ell)$$

$$= -4(g(X_i, X_j)g(X_k, X_\ell) - g(X_k, X_j)g(X_i, X_\ell)).$$

Using $\nabla_{\varphi X_i} X_j = 0$ and $[\varphi X_i, \varphi X_j] = 0$ we see that $g(R_{X_k X_i} \varphi X_j, \varphi X_\ell) = g(R_{\varphi X_j \varphi X_\ell} X_k, X_i) = 0$ and hence

$$g(R_{X_k X_i} X_j, X_\ell) = 4(g(X_i, X_j)g(X_k, X_\ell) - g(X_k, X_j)g(X_i, X_\ell))$$

completing the proof.

2. Flat Associated Metrics on $\mathbb{R}^3$ and other Metrics on $\mathbb{R}^{2n+1}$.

In dimension 3 it is easy to construct flat contact metric structures. For example, consider $\mathbb{R}^3$ with coordinates (x^1,x^2,x^3). In example I.2.C we noted that the 1-form $\eta = \frac{1}{2}(\cos x^3 dx^1 + \sin x^3 dx^2)$ is a contact form. In this case $\xi = 2(\cos x^3\ \partial/\partial x^1 + \sin x^3\ \partial/\partial x^2)$ and the metric g whose components are $g_{ij} = \frac{1}{4}\delta_{ij}$ gives a flat contact metric structure. Following the proof of the main theorem in section 1, we see that $\partial/\partial x^3$ spans the distribution $[+1]$ and $\sin x^3\ \partial/\partial x^1 - \cos x^3\ \partial/\partial x^2$ spans $[-1]$. Geometrically ξ is parallel along $[-1]$ and rotates (and hence the contact distribution D rotates) as we move parallel to the x^3-axis. Since η is invariant under the group of translations generated by $\{x^i \longrightarrow x^i + 2\pi,\ i = 1,2,3\}$, the 3-dimensional torus also carries this structure.

We can now find a flat associated metric on $\mathbb{R}^3$ for the standard contact form $\eta_o = \frac{1}{2}(dz - ydx)$. Consider the diffeomorphism $f\colon \mathbb{R}^3 \longrightarrow \mathbb{R}^3$ given by

$$x^1 = z\cos x - y\sin x,$$

$$x^2 = -z\sin x - y\cos x,$$

$$x^3 = -x.$$

Then $\eta_o = f^*\eta$ and the Riemannian metric $g_o = f^*g$ is a flat associated metric for the contact form η_o. The matrix of components of g_o is

$$\frac{1}{4}\begin{pmatrix} 1+y^2+z^2 & z & -y \\ z & 1 & 0 \\ -y & 0 & 1 \end{pmatrix}.$$

This metric generalizes to give an associated metric to the standard structure on $\mathbb{R}^{2n+1}$ other than the one usually studied (examples II.4.A, IV.6.A, V.2.B). At the same time this gives an example of the second theorem of section 1. Consider $\mathbb{R}^{2n+1}$ with coordinates $(x^i, y^i, z), i = 1, \cdots, n$, and $\eta = \frac{1}{2}(dz - \sum_{i=1}^{n} y^i dx^i)$, the standard contact form. ξ is $2\,\partial/\partial z$ and the Riemannian metric g with component matrix

$$\frac{1}{4}\begin{pmatrix} \delta_{ij} + y^i y^j + \delta_{ij}z^2 & \delta_{ij}z & -y^i \\ \delta_{ij}z & \delta_{ij} & 0 \\ -y^j & 0 & 1 \end{pmatrix}$$

is an associated metric for η. The matrix of components of φ is

$$\begin{pmatrix} \delta_{ij}z & \delta_{ij} & 0 \\ -\delta_{ij} - \delta_{ij}z^2 & -\delta_{ij}z & 0 \\ y^j z & y^j & 0 \end{pmatrix}.$$

Now

$$\frac{1}{2}(\mathfrak{L}_\xi\varphi)\,\partial/\partial y^i = [\partial/\partial z, \partial/\partial x^i - z\,\partial/\partial y^i + y^i\partial/\partial z]$$
$$= -\,\partial/\partial y^i\,.$$

Thus $\partial/\partial y^i$ is an eigenvector of the operator $h = \frac{1}{2}\mathfrak{L}_\xi\varphi$ with eigenvalue -1. Since φ and h anti-commute for any contact metric structure, $h\varphi\,\partial/\partial y^i = -\varphi h\partial/\partial y^i = \varphi\,\partial/\partial y^i$ so that $\varphi\,\partial/\partial y^i$ is an eigenvector of h with eigenvalue $+1$. Thus the contact distribution D is decomposed into the ± 1 eigenspaces of h. Again since $\nabla_X\xi = -\varphi hX - \varphi X$ for any contact metric structure, ξ is parallel along the distribution $[-1]$ which is spanned by the $\partial/\partial y^i$'s and we have $R_{\partial/\partial y^i\partial/\partial y^j}\xi = 0$. Recall that the integral curves of ξ are geodesics of an associated metric and hence we have $R_{\xi\partial/\partial y^i}\xi = 0$. Similarly since $[\xi, \varphi\,\partial/\partial y^i] = -\,2\,\partial/\partial y^i$,

$$R_{\xi\varphi\,\partial/\partial y^i}\,\xi = \nabla_\xi(-2\varphi^2\partial/\partial y^i) - \nabla_{-2\partial/\partial y^i}\xi$$
$$= 2\,\nabla_\xi\,\partial/\partial y^i$$
$$= 2\,\nabla_{\partial/\partial y^i}\xi$$
$$= 0\,.$$

Thus $R_{\xi X}\xi = 0$ for all X and $R_{XY}\xi = 0$ for X and Y belonging to $[-1]$. Note that $[+1]$ is spanned by the fields $\varphi\partial/\partial y^i = \partial/\partial x^i - z\partial/\partial y^i + y^i\partial/\partial z$ and hence is not integrable.

Comparing the two associated metrics for $\eta = \frac{1}{2}(dz - \sum_{i=1}^{n} y^i dx^i)$ on $\mathbb{R}^{2n+1}$, for the Sasakian one we $\nabla_X \xi = -\varphi X$ for all vector fields X and for the one given here $\nabla_X \xi = 0$ or $-2\varphi X$ according as X belongs to $[-1] \oplus [\xi]$ or $[+1]$ respectively. Geometrically in the first case the contact distribution D rotates whenever we move along an integral curve of D, whereas for the second metric all the rotation is in the n-dimensional distribution $[+1]$.

CHAPTER VII

THE TANGENT SPHERE BUNDLE

1. Differential Geometry of the Tangent Bundle.

Before turning to the tangent sphere bundle in section 2, we give the necessary preliminaries on the tangent bundle of a Riemannian manifold. More complete studies of the tangent bundle are given in Dombrowski [17], Sasaki [55,57] and Yano and Ishihara [86].

As in example I.2.E let M be an $(n+1)$-dimensional manifold and TM its tangent bundle with projection map $\bar{\pi}$. If $(x^1, \cdots, x^{n+1})$ are local coordinates on M, we set $q^i = x^i \circ \bar{\pi}$. Then $(q^1, \cdots, q^{n+1})$ together with the fibre coordinates $(v^1, \cdots, v^{n+1})$ form local coordinates on TM.

Let ω be a 1-form on M. Following Yano and Ishihara [86] we define a function $\gamma\omega$ on TM by

$$(\gamma\omega)(Z) = \omega(Z),$$

$Z \in TM$. If now D is a derivation on M, i.e. a derivation on the tensor algebra on M, we define the <u>lift</u> of D to TM to be the vector field $\bar{D}$ on TM such that

$$\bar{D}df = \gamma(Ddf)$$

where for a function f on M, on the right side D is acting on the 1-form df and on the left side df is regarded as a function on TM.

For a symmetric connection D on M and a vector field X on M we define the horizontal lift X^H of X to TM to be the lift of the derivation D_X. If Γ_{ij}^k are the connection coefficients of D then the local expression for X^H is

$$X^H = X^i \frac{\partial}{\partial q^i} - X^i v^j \Gamma_{ij}^k \frac{\partial}{\partial v^k} \tag{1}$$

where the X^i's are the components of X with respect to the local coordinates $(x^1, \cdots, x^{n+1})$.

On the other hand the vertical lift X^V of a vector field X is simply defined by

$$X^V \omega = \omega(X) \circ \bar{\pi}$$

where again on the left side the 1-form ω is regarded as a function on TM.

For functions the vertical and horizontal lifts are defined by $f^V = f \circ \bar{\pi}$ and $f^H = 0$. Moreover note that $(fX)^H = f^V X^H$.

We now define the connection map K of D [17] mapping TTM onto TM. Let Z be a point in TM, then K is defined by

$$KX^H = 0, \quad K(X^V{}_Z) = X_{\bar{\pi}(Z)} .$$

Similarly we define an almost complex structure J on TM by

$$JX^H = X^V, \quad JX^V = -X^H .$$

Dombrowski [17] shows that since D is symmetric J is integrable if and only if D has vanishing curvature $\underline{R}$ on M (in fact defining the horizontal lift for a non-symmetric connection, the integrability of J is equivalent to the vanishing of both the curvature and torsion of the connection [17]).

If now G is a Riemannian metric on M and D its Riemannian connection, we define a Riemannian metric $\bar{g}$ on TM, called the <u>Sasaki metric</u> [55], by

$$\bar{g}(X,Y) = G(\bar{\pi}_* X, \bar{\pi}_* Y) + G(KX,KY)$$

where here X and Y are vectors on TM. $\bar{g}$ is a Hermitian metric for the almost complex structure J on TM. Indeed, since $\bar{\pi}_* \circ J = -K$ and $K \circ J = \bar{\pi}_*$,

$$\begin{aligned} \bar{g}(JX,JY) &= G(\bar{\pi}_* JX, \bar{\pi}_* JY) + G(KJX,KJY) \\ &= G(KX,KY) + G(\bar{\pi}_* X, \bar{\pi}_* Y) \\ &= \bar{g}(X,Y), \end{aligned}$$

for vectors X and Y on TM.

In example I.2.E, we defined a 1-form β on TM by the local expression $\beta = \sum_{i,j} G_{ij} v^j dq^i$; equivalently it is given by

$$\beta(X)_Z = G(Z, \bar{\pi}_* X)$$

where $X \in T_Z TM$. Moreover $2d\beta$ is the fundamental 2-form of the almost Hermitian structure $(J,\bar{g})$ on TM [17]. Thus we see that $(J,\bar{g})$ is an almost Kähler structure on TM which is Kählerian if and only if the Riemannian metric G on M is flat (Dombrowski [17], Tachibana and Okumura [68]).

Finally the Riemannian connection $\bar{\nabla}$ of the metric $\bar{g}$ on TM is given at $Z \in TM$ by

$$(\bar{\nabla}_{X^H} Y^H)_Z = (D_X Y)^H{}_Z - \frac{1}{2}(\underline{R}_{XY} Z)^V$$
$$(\bar{\nabla}_{X^H} Y^V)_Z = -\frac{1}{2}(\underline{R}_{YZ} X)^H + (D_X Y)^V{}_Z \qquad (2)$$
$$(\bar{\nabla}_{X^V} Y^H)_Z = -\frac{1}{2}(\underline{R}_{XZ} Y)^H$$
$$\bar{\nabla}_{X^V} Y^V = 0 .$$

The curvature tensor of $\bar{\nabla}$ will be denoted by $\bar{R}$.

2. Geometry of the Tangent Sphere Bundle as a Contact Metric Manifold.

We have seen that the principal circle bundles form a large class of examples of contact manifolds; they have K-contact structures and Sasakian structures if the base manifolds are Kählerian. Thus these examples together with the usual contact metric structure on $\mathbb{R}^{2n+1}$ (examples II.4.A and IV.6.A) show that Sasakian manifolds form a large and important class of contact manifolds. The tangent and cotangent sphere bundles however are in general not of this type as we shall see in this section, even though they are classically an important class of examples of contact manifolds.

The tangent sphere bundles as contact manifolds have not been widely studied from the Riemannian point of view, so we will briefly present the geometry of a contact metric structure on the tangent sphere bundle. As is customary we regard the tangent sphere bundle as the bundle of unit tangent vectors, even though owing to the factor $\frac{1}{2}$ in the coboundary formula for $d\eta$ a homothetic change of metric will be made. (If one adopts the convention that the $\frac{1}{2}$ does not appear in the coboundary formula for $d\eta$, this change is not necessary. However to be consistent the odd-dimensional sphere as a standard example of a Sasakian manifold should then be taken as a sphere of radius 2. (Compare Tashiro [77] and Sasaki and Hatakeyama [61])).

The tangent sphere bundle $\pi\colon T_1M \to M$ is a hypersurface of TM given by $\sum(v^i)^2 = 1$. Note that $\pi = \bar{\pi} \circ \iota$ where ι is the immersion. The vector field $N = v^i\partial/\partial v^i$ is a unit normal as well as the position vector for a point Z of T_1M. We denote by g' the induced metric $\iota^*\bar{g}$ and by ∇ its Riemannian connection. We can easily find the Weingarten map of the immersion. For any vertical tangent vector field U,

$$\bar{\nabla}_{\iota_*U}N = (\iota_*Uv^i)\frac{\partial}{\partial v^i} + v^i\bar{\nabla}_{\iota_*U}\left(\frac{\partial}{\partial x^i}\right)^V$$
$$= \iota_*U\,.$$

For a horizontal tangent vector field X, we may suppose that ι_*X is (the restriction of) a horizontal lift, then

$$(\bar{\nabla}_{(\partial/\partial x^j)^H}N)_Z = \left(\left(\frac{\partial}{\partial x^j}\right)^H v^i\right)\frac{\partial}{\partial v^i}_Z - \frac{1}{2}v^i(R_{\partial/\partial x^i Z}\,\partial/\partial x^j)^H$$
$$+ v^i(D_{\partial/\partial x^j}\partial/\partial x^i)^V_Z$$
$$= 0$$

by equation (1). Thus the Weingarten map H is given by $HU = -U$ for any vertical tangent vector U and $HX = 0$ for any horizontal vector X.

We know that as a hypersurface of the almost Kähler manifold TM, T_1M inherits an almost contact metric structure. Following the usual procedure (example II.4.B) we define φ', ξ' and η' by $\iota_*\xi' = -JN = -v^iJ\left(\frac{\partial}{\partial x^i}\right)^V = v^i\left(\frac{\partial}{\partial x^i}\right)^H$ and

$J\iota_* X = \iota_* \varphi' X + \eta'(X) N$. $(\varphi', \xi', \eta', g')$ is then an almost contact metric structure. Moreover η' is the contact form on T_1M induced from the 1-form β on TM, for

$$\begin{aligned}\eta'(X) &= \bar{g}(N, J\iota_* X) \circ \iota = 2d\beta(N, \iota_* X) \circ \iota \\ &= 2\sum (d(G_{ij}v^j) \wedge dq^i)(v^k \frac{\partial}{dv^k}, \iota_* X) \circ \iota \\ &= \sum G_{ik} v^k dq^i (\iota_* X) \circ \iota \\ &= \iota^* \beta(X).\end{aligned}$$

However $g'(X, \varphi' Y) = 2d\eta'(X, Y)$, so that $(\varphi', \xi', \eta', g')$ is not a contact metric structure. Of course, the difficulty is easily rectified and we shall take

$$\eta = \frac{1}{2}\eta' = \iota^*(\frac{1}{2}\beta), \xi = 2\xi' = 2v^i (\frac{\partial}{\partial x^i})^H, \quad \varphi = \varphi', g = \frac{1}{4}g'$$

as our contact metric structure on T_1M.

Before proceeding to our theorems we shall obtain explicitly the covariant derivatives of ξ and φ. For a horizontal tangent vector field X we again take

$\iota_* X = (\frac{\partial}{\partial x^j})^H$, then

$$\begin{aligned}(\iota_* \nabla_X \xi)_Z &= (\bar{\nabla}_{\iota_* X} \iota_* \xi)_Z \\ &= ((\frac{\partial}{\partial x^j})^H 2v^i)(\frac{\partial}{\partial x^i})^H_Z + 2v^i (D_{\partial/\partial x^j} \partial/\partial x^i)^H_Z \\ &\quad - v^i (\underline{R}_{\partial/\partial x^j \partial/\partial x^i} Z)^V \\ &= -(\underline{R}_{\pi_* X Z} Z)^V \qquad (3)\end{aligned}$$

For a vertical tangent vector field U we have

$$(\iota_*\nabla_U\xi)_Z = (\bar{\nabla}_{\iota_*U}\iota_*\xi)_Z$$

$$= (\iota_*U2v^i)\left(\frac{\partial}{\partial x^i}\right)^H_Z - v^i(\underline{R}_{K\iota_*UZ}\ \partial/\partial x^i)^H$$

$$= -\ 2\iota_*\varphi U_Z - (\underline{R}_{K\iota_*UZ}Z)^H \qquad (4)$$

since $\left(\frac{\partial}{\partial x^i}\right)^H = -\ J\left(\frac{\partial}{\partial x^i}\right)^V$.

Now to differentiate φ, first note that for any tangent vector fields X and Y

$$\iota_*(\nabla_X\varphi)Y = \bar{\nabla}_{\iota_*X}J\iota_*Y - (\nabla_X\eta')(Y)N + \eta'(Y)\iota_*HX$$
$$-\ h(X,\varphi Y)N - J\bar{\nabla}_{\iota_*X}\iota_*Y - h(X,Y)\iota_*\xi'$$

where h is the second fundamental form. Again for X and Y horizontal we suppose that they are horizontal lifts and we let $U = U^i\frac{\partial}{\partial v^i}$ and $W = W^i\frac{\partial}{\partial v^i}$ be vertical tangent vector fields.

$$(\iota_*(\nabla_X\varphi)Y)_Z = -\ \frac{1}{2}(\underline{R}_{\pi_*YZ}\ \pi_*X)^H + (D_{\pi_*X}\pi_*Y)^V_Z - (\nabla_X\eta')(Y)N_Z$$
$$-\ J(D_{\pi_*X}\pi_*Y)^H_Z + \frac{1}{2}\ J(\underline{R}_{\pi_*X\pi_*Y}Z)^V$$
$$= -\ \frac{1}{2}(\underline{R}_{\pi_*XZ}\pi_*Y)^H \qquad (5)$$

$$(\iota_*(\nabla_X\varphi)U)_Z = -(\iota_*XU^i)\left(\frac{\partial}{\partial x^i}\right)^H_Z - U^i(D_{\pi_*X}\partial/\partial x^i)^H_Z$$

$$+ \frac{1}{2}(\underline{R}_{\pi_*X K\iota_*U}Z)^V - J(\iota_*XU^i)\left(\frac{\partial}{\partial x^i}\right)^V_Z$$

$$- U^iJ(D_{\pi_*X}\partial/\partial x^i)^V_Z + \frac{1}{2}J(\underline{R}_{K\iota_*UZ}\pi_*X)^H$$

$$- (\nabla_X\eta')(U)N_Z$$

$$= \frac{1}{2}\tan\,(\underline{R}_{\pi_*XZ}K\iota_*U)^V \qquad (6)$$

where tan denotes the tangential part.

$$(\iota_*(\nabla_U\varphi)X)_Z = -\;2\eta(X)\,\iota_*U_Z + \frac{1}{2}J(\underline{R}_{K\iota_*UZ}\pi_*X)^H$$

$$- h(U,\varphi X)N_Z - (\nabla_U\eta')(X)N_Z$$

$$= -\;2\eta(X)\,\iota_*U_Z + \frac{1}{2}\tan\,(\underline{R}_{K\iota_*UZ}\pi_*X)^V \qquad (7)$$

$$(\iota_*(\nabla_U\varphi)W)_Z = -\;(\iota_*UW^i)\left(\frac{\partial}{\partial x^i}\right)^H_Z + \frac{1}{2}(\underline{R}_{K\iota_*UZ}K\iota_*W)^H$$

$$- (\nabla_U\eta')(W)N_Z - J(\iota_*UW^i)\left(\frac{\partial}{\partial x^i}\right)^V_Z + g'(U,W)\,\iota_*\xi'_Z$$

$$= 2g(U,W)\,\iota_*\xi_Z + \frac{1}{2}(\underline{R}_{K\iota_*UZ}K\iota_*W)^H \qquad (8)$$

We are now in a position to show that the contact metric structure on T_1M is not in general K-contact; in fact only rarely. The following theorem is due to Y. Tashiro [77].

<u>Theorem</u>. The contact metric structure (φ,ξ,η,g) on T_1M is K-contact if and only if M has positive constant curvature 1 in which case T_1M is a Sasakian manifold.

<u>Proof</u>. If (φ,ξ,η,g) is a K-contact structure, then $\nabla_X\xi = -\varphi X$ as we have seen. Taking X such that ι_*X is a horizontal lift, equation (3) gives $(\iota_*\varphi X)_Z = (\underline{R}_{\pi_*XZ}Z)^V$. Thus if we take an orthonormal pair of vectors <u>on</u> M, <u>say</u> X,Z, $\underline{R}_{XZ}Z = X$ from which $\underline{R}_{XY}Z = G(Y,Z)X - G(X,Z)Y$.

Conversely if M has constant curvature equal to 1, we claim that T_1M is Sasakian. Since $\underline{R}_{XY}Z = G(Y,Z)X - G(X,Z)Y$ on M equations (5) through (8) on T_1M become

$$(\iota_*(\nabla_X\varphi)Y)_Z = -\frac{1}{2}(G(\pi_*Y,Z)\pi_*X - G(\pi_*X,\pi_*Y)Z)^H$$

$$= \iota_*(g(X,Y)\xi - \eta(Y)X)_Z,$$

$$(\iota_*(\nabla_X\varphi)U)_Z = \frac{1}{2}\tan(G(K\iota_*U,Z)\pi_*X - G(\pi_*X,K\iota_*U)Z)^V$$

$$= 0,$$

$$(\iota_*(\nabla_U\varphi)X)_Z = -2\eta(X)\,\iota_*U_Z$$

$$+\frac{1}{2}\tan\,(G(Z,\pi_*X)K\iota_*U - G(K\iota_*U,\pi_*X)Z)^V$$

$$= -2\eta(X)\,\iota_*U_Z + \frac{1}{2}(2\eta(X)\,\iota_*U_Z)$$

$$= \iota_*(\eta(X)U)_Z,$$

$$(\iota_*(\nabla_U\varphi)W)_Z = 2g(U,W)\,\iota_*\xi_Z$$

$$+ \frac{1}{2}(G(Z,K\iota_*W)K\iota_*U - G(K\iota_*U,K\iota_*W)Z)^H$$

$$= \iota_*(g(U,W)\xi)_Z .$$

In Chapter VI we saw that if on a contact metric manifold M^{2n+1} $R_{XY}\xi = 0$, then the manifold is locally the product of a flat (n+1)-dimensional manifold and an n-dimensional manifold of positive constant curvature 4. Here we give an example of such a contact manifold [7].

<u>Theorem</u>. The contact metric structure (φ,ξ,η,g) on T_1M satisfies $R_{\xi U}\xi = 0$ for all vertical vector fields U if and only if M is flat in which case $R_{XY}\xi = 0$ for all vector fields X and Y on T_1M.

<u>Proof</u>. From equations (2) we have

$$K\bar{R}_{X^H U^V}Y^H = \frac{1}{4}\underline{R}_{X\underline{R}_{UZ}Y}Z + \frac{1}{2}\underline{R}_{XY}U$$

for any three vector fields X,U and Y on M (see also Kowalski [34]). If now we let U be a vertical tangent vector field on T_1M, then $R_{\xi U}\xi = 0$ implies that

$$\underline{R}_{Z\underline{R}_{K\iota_*UZ}Z}Z = 0$$

and hence that

$$\underline{R}_{Z\underline{R}_{XZ}Z}Z = 0$$

for all vectors X and Z on M. Therefore

$$0 = G(\underline{R}_{Z\underline{R}_{ZX}Z}Z,X) = G(\underline{R}_{ZX}Z,\underline{R}_{ZX}Z) ,$$

that is $\underline{R}_{ZX}Z = 0$ on M. Linearizing this and using the Bianchi identity we have that $\underline{R}_{XY}Z = 0$ for all X,Y,Z on M.

Conversely if M is flat, equations (3) and (4) give $\nabla_X\xi = 0$ for X horizontal and $\nabla_U\xi = -2\varphi U$ for U vertical. Thus the vertical distribution on T_1M is the [+1] distribution of Chapter VI and the horizontal distribution is the $[-1] \oplus [\xi]$ distribution. Therefore [+1] is integrable and since M is flat $[-1] \oplus [\xi]$ is also integrable. Thus for X and Y horizontal on T_1M and U and W vertical we may take these as coordinate vectors as in Chapter VI. Now $R_{XY}\xi = 0$ is trivial,

$$R_{XU}\xi = -2\nabla_X\varphi U = 0$$

by equation (2) of Chapter VI and

$$\begin{aligned} R_{UW}\xi &= -2\nabla_U\varphi W + 2\nabla_W\varphi U \\ &= 0 \end{aligned}$$

by equation (3) of Chapter VI .

REFERENCES

1. Abe, K., Some examples of non-regular almost contact structure on exotic spheres, to appear.

2. Abe, K. and J. Erbacher, Non-regular contact structures on Brieskorn manifolds, Bull. AMS 81(1975) 407-409.

3. Andreotti, A. and C.D. Hill, Complex characteristic coordinates and tangential Cauchy-Riemann equations, Ann. Sc. Norm. Sup. Pisa, 26(1972) 299-324.

4. Blair, D.E., The theory of quasi-Sasakian structures, J. Diff. Geom. 1(1967) 331-345.

5. Blair, D.E., Almost contact manifolds with Killing structure tensors, Pacific J. of Math. 39(1971) 285-292.

6. Blair, D.E., On the non-existence of flat contact metric structures, to appear.

7. Blair, D.E., Two remarks on contact metric structures, to appear.

8. Blair, D.E. and S.I. Goldberg, Topology of almost contact manifolds, J. Diff. Geom. 1(1967) 347-354.

9. Blair, D.E. and K. Ogiue, Geometry of integral submanifolds of a contact distribution, Illinois J. Math., 19(1975) 269-276.

10. Blair, D.E. and K. Oguie, Positively curved integral submanifolds of a contact distribution, to appear.

11. Blair, D.E. and D.K. Showers, Almost contact manifolds with Killing structure tensors, II, J. Diff. Geom. 9(1974) 577-582.

12. Blair, D.E., D.K. Showers and K. Yano, Nearly Sasakian structures, to appear.

13. Boothby, W.M. and H.C. Wang, On contact manifolds, Ann. of Math., 68(1958) 721-734.

14. Cartan, E., Lecons sur les Invariants Intégraux, Hermann Paris, 1922.

15. Chern, S.S., Pseudo-groupes continus infinis, colloques Internationaux du C.N.R.S., Strasbourg, 1953, 119-136.

16. Chern, S.S., M.P. do Carmo and S. Kobayashi, Minimal submanifolds of a sphere with second fundamental form of constant length, Functional Analysis and Related Fields, Springer-Verlag, 1970, 59-75.

17. Dombrowski, P., On the geometry of the tangent bundle, J. reine und angew. Math., 210(1962) 73-88.

18. Fujitani, T., Complex-valued differential forms on normal contact Riemannian manifolds, Tôhoku Math. J. 18(1966) 349-361.

19. Goldberg, S.I., Curvature and Homology, Academic Press, New York, 1962.

20. Goldberg, S.I., Rigidity of positively curved contact manifolds, J. London Math. Soc. 42(1967) 257-263.

21. Goldberg, S.I., Totally geodesic hypersurfaces of Kaehler manifolds, Pacific J. of Math. 27(1968) 275-281.

22. Goldberg, S.I., On the topology of compact contact manifolds, Tôhoku Math. J. 20(1968) 106-110.

23. Gray, J., Some global properties of contact structures, Ann. of Math., 69(1959), 421-450.

24. Harada, M., On the curvature of Sasakian manifolds, Bull. Yamagata Univ. Nat. Sci. 7(1969) 97-106.

25. Harada, M., On the minimal diameter of Sasakian manifolds Bull. Yamagata Univ. Nat. Sci. 7(1970) 191-203.

26. Hatakeyama, Y., Some notes on differentiable manifolds with almost contact structures, Tôhoku Math. J., 15(1963) 176-181.

27. Hatakeyama, Y., Y. Ogawa and S. Tanno, Some properties of manifolds with contact metric structures, Tôhoku Math J., 15(1963) 42-48.

28. Huygens, C., Traité de la Lumiere, Vander Aa, Leiden, 1690.

29. Ianus, S., Sulle varietà di Cauchy-Riemann, Rend. dell'Accademia di Scienze Fisiche e Matemtiche, Napoli, XXXIX(1972) 191-195.

30. Kleinfield, E., A characterization of the Cayley numbers, Studies in Modern Algebra, MAA Studies in Mathematics, Vol. 2, 1963, 126-143.

31. Kobayashi, S., Principal fibre bundles with 1-dimensional toroidal group, Tôhoku Math. J., 8(1956), 29-45.

32. Kobayashi, S., Transformation Groups in Differential Geometry, Springer-Verlag, 1972.

33. Kobayashi, S. and K. Nomizu, Foundations of Differential Geometry, Vol. 2, John Wiley and Sons, New York, 1969.

34. Kowalski, O., Curvature of the induced Riemannian metric on the tangent bundle of a Riemannian manifold, J. reine und angew.Math., 250(1971) 124-129.

35. Kurita, M., On normal contact metric manifolds, J. Math. Soc. Japan, 15(1963) 304-318.

36. Libermann, P., Sur les automorphismes infinitesimaux des structures symplectiques et des structures de contact, Colloq. Géométrie Différentielle Globale (Bruxelles, 1958), Louvain, 1959, 37-59.

37. Lie, S., Theorie der Transformationgruppen, Vol. 2, Leipzig, Teubner, 1890.

38. Ludden, G.D., Submanifolds of manifolds with an f-structure, Kōdai Math. Sem. Rep., 21(1969) 160-166

39. MacLane, S., Geometrical Mechanics II, Lecture notes, University of Chicago, 1968.

40. Martinet, J., Formes de contact sur les variétiés de dimension 3, Proc. of Liverpool Singularities Symposium II, Springer-Varlag, 1971, 142-163.

41. Morimoto, A., On normal almost contact structures, J. Math. Soc. Japan 15(1963) 420-436.

42. Morimoto, A., On normal almost contact structure with a regularity, Tôhoku Math. J. 16(1964) 90-104.

43. Moskal, E.M., Contact manifolds of positive curvature, Thesis, University of Illinois, 1966.

44. Newlander, A. and L. Nirenberg, Complex analytic coordinates in almost complex manifolds, Ann. of Math., 65(1957) 391-404.

45. Ogiue, K., On almost contact manifolds admitting axiom of planes or axiom of free mobility, Kōdai Math. Sem. Rep. 16(1964) 223-232.

46. Ogiue, K., On fiberings of almost contact manifolds, Kōdai Math. Sem. Rep. 17(1965) 53-62.

47. Okumura, M., Some remarks on space with a certain contact structure, Tôhoku Math. J. 14(1962) 135-145.

48. Okumura, M., On infinitesimal conformal and projective transformations of normal contact spaces, Tôhoku Math. J. 14(1962) 398-412.

49. Okumura, M., Certain almost contact hypersurfaces in Euclidean spaces, Kōdai Math. Sem. Rep. 16(1964) 44-54.

50. Okumura, M., Certain almost contact hypersurfaces in Kaehlerian manifolds of constant holomorphic sectional curvatures, Tôhoku Math. J. 16(1964) 270-284.

51. Okumura, M., Cosymplectic hypersurfaces in Kählerian manifold of constant holomorphic sectional curvature, Kōdai Math. Sem. Rep. 17(1965) 63-73.

52. Okumura, M., Contact hypersurfaces in certain Kaehlerian manifolds, Tôhoku Math. J. 18(1966) 74-102.

53. Porteous, I.R., Topological Geometry, Van Nostrand Reinhold, London, 1969.

54. Reeb, G., Sur certaines propriétés topologiques des trajectoires des systemes dynamiques, Mémoires des l'Acad. Roy. de Belgique, Sci. Ser. 2, 27(1952) 1-62.

55. Sasaki, S., On the differential geometry of tangent bundles of Riemannian manifolds, Tôhoku Math. J., 10(1958) 338-354.

56. Sasaki, S., On differentiable manifolds with certain structures which are closely related to almost contact structure I, Tôhoku Math. J., 12(1960) 459-476.

57. Sasaki, S., On the differential geometry of tangent bundles of Riemannian manifolds II, Tôhoku Math. J. 14(1962) 146-155.

58. Sasaki, S., A characterization of contact transformations, Tôhoku Math. J. 16(1964) 285-290.

59. Sasaki, S., Almost Contact Manifolds, Lecture Notes, Mathematical Institute, Tôhoku University, Vol. 1, 1965, vol. 2, 1967, vol. 3, 1968.

60. Sasaki, S. and Y. Hatakeyama, On differentiable manifolds with certain structures which are closely related to almost contact structure II, Tôhoku Math. J., 13(1961) 281-294.

61. Sasaki, S. and Y. Hatakeyama, On differentiable manifolds with contact metric structures, J. Math. Soc. Japan, 14(1962) 249-271.

62. Sasaki, S. and C.J. Hsu, On the integrability of almost contact structure, Tôhoku Math. J., 14(1962) 167-176.

63. Sato, I., On a structure similar to Sasakian 3-structure, Tôhoku Math. J. 25(1973) 405-415.

64. Steenrod, N., Topology of Fibre Bundles, Princeton University Press, Princeton, 1951.

65. Sternberg, S., Lectures on Differential Geometry, Prentice Hall, Englewood Cliffs, 1964.

66. Stong, R.E., Contact manifolds, J. Diff. Geom. 9(1974) 219-238.

67. Tachibana, S., On harmonic tensors in compact Sasakian spaces, Tôhoku Math. J., 17(1965) 271-284.

68. Tachibana, S. and M. Okumura, On the almost complex structure of tangent bundles of Riemannian spaces, Tôhoku Math. J., 14(1962) 156-161.

69. Takizawa, S., On contact structures of real and complex manifolds, Tôhoku Math. J., 15(1963) 227-252.

70. Tanno, S., A theorem on regular vector fields and its applications to almost contact structures, Tôhoku Math. J. 17(1965) 235-238.

71. Tanno, S., Harmonic forms and Betti numbers of certain contact Riemannian manifolds, J. Math. Soc. Japan 19(1967) 308-316.

72. Tanno, S., The topology of contact Riemannian manifolds, Illinois J. Math., 12(1968) 700-717.

73. Tanno, S., The automorphism groups of almost contact Riemannian manifolds, Tôhoku Math. J. 21(1969) 21-38.

74. Tanno, S., Sasakian manifolds with constant φ-holomorphic sectional curvature, Tôhoku Math. J., 21(1969) 501-507.

75. Tashiro, Y., On contact structures of hypersurfaces in almost complex manifolds I., Tôhoku Math. J. 15(1963) 62-78.

76. Tashiro, Y., On contact structure of hypersurfaces in complex manifolds II, Tôhoku Math. J. 15(1963) 167-175.

77. Tashiro, Y., On contact structures of tangent sphere bundles, Tôhoku Math. J., 21(1969) 117-143.

78. Thomas, C.B., Almost regular contact manifolds, to appear.

79. Yamaguchi, S. and T. Ikawa, On compact minimal C-totally real submanifolds, Tensor N.S., 26(1975) 9-16.

80. Yamaguchi, S., M. Kon and T. Ikawa, On C-totally real submanifolds, J. Diff. Geom., to appear.

81. Yamaguchi, S., M. Kon and Y. Miyahara, A theorem on C-totally real minimal surfaces, Proc. AMS, to appear.

82. Yano, K., On a structure defined by a tensor field of type (1,1) satisfying $f^3 + f = 0$, tensor, N.S., 14(1963) 9-19.

83. Yano, K., Differential Geometry on Complex and Almost Complex Spaces, Pergamon, New York, 1965.

84. Yano, K. and S. Ishihara, On integrability of a structure f satisfying $f^3 + f = 0$, Quart. J. Math. Oxford (2), 15(1964) 217-222.

85. Yano, K. and S. Ishihara, The f-structure induced on submanifolds of complex and almost complex spaces, Kōdai Math. Sem. Rep., 18(1966) 120-160.

86. Yano, K. and S. Ishihara, Tangent and Cotangent Bundles, Dekker, New York, 1973.

INDEX